# WIRELESS SENSOR NETWORKS - TECHNOLOGY AND APPLICATIONS

# WIRELESS SENSOR NETWORKS - TECHNOLOGY AND APPLICATIONS

*Editor*

**Carlos Granger**

scitus
academics

**Wireless Sensor Networks - Technology and Applications**
Edited by **Carlos Granger**

ISBN: 978-1-68117-611-6
Library of Congress Control Number: 2016946466


SCITUS Academics LLC,
616, Corporate Way, Suite 2, 4766,
Valley Cottage, NY 10989

www.scitusacademics.com



**Notice**



Printed in United States of America on Acid Free Paper

# PREFACE

Also developments in wireless communications and wireless sensor networks (WSNs) have emerged with the development of microelectromechanical system technology (MEMS). In recent years, WSNs have become probably the most fascinating field of investigation. A WSN consists of a range of wireless sensor nodes, which together with a sink form a sensor field. These many nodes, which have the ability to feel their surroundings, carry out limited computations and communicate wirelessly with the WSN. WSNs can be found in a variety of civil and military applications worldwide, such as detection of enemy battlefield intrusion, habitat monitoring, object tracking, patient monitoring and fire detection. Sensor networks emerge as an attractive technology with a strong future promise. However, challenges in deployment and coverage, computational power, size, quality of service, scalability, security and energy efficiency remain to be addressed.

Wireless Sensor Networks - Technology and Applications presents important issues for WSNs. The book is an important tool for senior graduates and scholars who want to learn about the latest developments in wireless sensor networks.

# TABLE OF CONTENTS

1

# EFFICIENT NODE PLACEMENT FOR CONGESTION CONTROL IN WIRELESS SENSOR NETWORKS

**Charalambos Sergiou[1] and Vasos Vassiliou[1]**

[1] University of Cyprus, 1 University Ave., 2019, Nicosia, Cyprus

## 1. INTRODUCTION

Wireless sensor nodes are small, embedded computing devices that interface with sensors/ actuators and communicate using short-range wireless transmitters. Such nodes act autonomously, but cooperatively to form a logical network, in which data packets are routed hop-by-hop towards management nodes, typically called sinks or base stations. A Wireless Sensor Network (WSN) comprises of a potentially large set of nodes that may be distributed over a wide geographical area, indoor or outdoor. Wireless Sensor Networks (WSNs) enable numerous sensing and monitoring services in areas of vital importance such as efficient industry production, safety and security at home as well as in traffic and environmental monitoring. Traffic patterns in WSNs can be derived from the physical processes that they sense. WSNs typically operate under light load and suddenly become active in response to a detected or monitored event. Early research studies in WSNs targeted military applications, especially for battlefield monitoring. In the last few years, due to the progress of low powered units and improvements in radio technologies, wireless sensor networks technologies have gained momentum. WSNs are now being deployed in civilian areas and being used for habitat observation ([1], [2]), health monitoring ([3]), object tracking ([4], [5]) etc. In addition, lately, there is an emergence of mission-critical applications ([6]).

Emergence of mission-critical and information demanding applications in WSNs renders performance control essential, for mission accomplishment. Heavy traffic load is a major factor that affects the performance of any type of network. The situation worsens in low-powered, unreliable WSNs. Thus, a prominent factor that under specific circumstances, can improve or deteriorate the performance of WSNs, can be the way that nodes are placed on the monitored field. Proper node placement is essential to ensure good sensing coverage and communication connectivity.

In this work we present and analyze four general ways that nodes can be placed on a grid and we compare the performance of a representative number of routing and congestion control algorithms under these placements.

The examined algorithms are ESRT ([7]), Sen- TCP ([8]), Directed Diffusion ([9]), HTAP ([10]) and Flooding. These algorithms are evaluated under four different placements. Simple Diffusion, Random, Grid and Biased- Random. Algorithms and placements are described in detail, in the next sections.

Initial results of this work have been presented in ([11] and [12]). In this paper we extent the number of evaluated algorithms in order to present a complete and solid work. Thus, we include an algorithm that represents the category of "reliable data transport" (ESRT), as well as a generic routing algorithm ("flooding"). Hence, in this paper, we provide representative results from all the categories of congestion control and reliable data transport algorithms in WSNs, under different placements.

Simulation results show that the performance of specific algorithms can be improved under specific placements. In particular, algorithms that employ multiple or alternative paths for forwarding the excess traffic from source to sink are favored by specific placements.

## 2. RELATED WORK

Several node placements have been proposed in literature concerning WSNs.

Younis et al ([13]) present a survey for strategies and techniques for node placements in WSNs and provide a categorization of the placement strategies into static and dynamic, depending on whether the optimization is performed at the time of deployment or while the network is operational.

Toumpis et al ([14]) provide an optimal deployment of large wireless sensor networks so as to minimize the number of nodes that is needed in order to transmit data from multiple sources to multiple sinks.

In ([15]) authors evaluate the tolerance against both random failure and battery exhaustion from the viewpoint of stochastic node placement. They consider three typical types of stochastic sensor placement: Simple diffusion, Constant Placement and R- Random placement.

In ([16]) authors studied the problem of determining the critical node density for maintaining k-coverage of a given square region. They have considered three different deployment strategies: Poisson point process, uniform random distribution and grid deployment and have shown that the two random strategies have identical density requirements for k-coverage. They also showed that grid deployment requires less node density than the two random deployments strategies in order to achieve the same level of coverage degree.

In ([11]) authors present a performance study for congestion control between three different algorithms under different node placements. Algorithms that employ three different techniques for congestion mitigation in WSNs. "Traffic control", "resource control" and multipath routing. Simulation results show that the performance of "resource control" algorithms is affected by different node placements.

In ([12]) authors study the energy utilization performance of HTAP algorithm([10]) under specific node placements, in correlation with Directed Diffusion ([9]) algorithm. Simulations results show that the performance of HTAP, a "resource control" algorithm, is improved when nodes are densely deployed near hotspots.

## 3. CONGESTION CONTROL IN WSNS

Congestion in WSNs occurs when the offered load is temporarily higher than the load which node(s) resources can process in a certain amount of time.

Congestion in WSNs can be categorized in two types. The first type of congestion happens in the medium. In this type, two or more nodes attempt to transmit simultaneously and as a result collisions of packets occur in the medium. This type of congestion is normally faced through enhancements in the MAC layer (e.g phase shifting that appeared in ([17]).

The other type of congestion happens when the queue or the buffer of a node used to hold packets to be transmitted, overflows. In such case packets drops happen, which is a highly undesirable situation in low powered WSNs. Solutions for this type of congestion lie in upper layers, like network or transport layer.

Generally, congestion in WSNs is mitigated by three categories of algorithms. "Traffic Control", "resource control" and "reliable data transport". "Traffic control" algorithms, affect the data rate of source nodes in order to reduce the traffic in the network when congestion occurs. Algorithms that employ this method, attempt, normally usually backpressure messages, to inform sources to reduce their data rate, in order to absorb the already high load and avoid packet drops.

On the other hand, "resource control" algorithms employ a different method in order to mitigate congestion. In this case, these algorithms attempt to take advantage of the already dense placement of WSNs, as well as the plethora of nodes that are in sleep state, by creating alternative or multiple paths to the sinks, in order to route the excess data. This type of algorithms do not affect the rate with which sources inject traffic in the network.

Finally, a different category is "reliable data transport" algorithms. This type of algorithms, typically run on the transport layer and focus on reliability. Although they are not "pure" congestion control algorithms, they can be considered as so, since congestion is a condition that affects significantly the reliability of WSNs.

Besides these categories there is another type of algorithms that attempts to create multiple paths in order to ease the transportation of data from source to sink. Although algorithms that fall in this category, cannot be considered as congestion control algorithms, we study them, since multiple paths can assist the network to balance the load and avoid congestion occurrence.

Thus, in this work we study the behavior of a representative algorithm of each category when nodes are placed under different placements.

Specifically we employ SenTCP ([8]) as a "traffic control" algorithm, HTAP ([10]) as a "resource control", ESRT [7]) as a "reliable data transport" algorithm as well as "Directed Diffusion" ([9]) as a "multiple path creation" algorithm. Furthermore, we also study "flooding algorithm" which is a generic routing algorithm, for comparison purposes.

# 4. NODE PLACEMENTS ANALYSIS

It is clear that node density is only one factor that affects network topology. The actual placement of nodes is also significant, as it is shown in ([13] and [18]). The placement of nodes affects the ability of a network to correctly sense an event while it also affects the number of possible disjoint paths towards the sink(s).

Thus, we claim that the placement of sensor nodes on a monitored field, is a factor that it is possible to affect the overall performance of the network.

Placement of nodes in a network can be divided into three major categories concerning the way that nodes are placed in the field. These are the deterministic node placement, the semi- deterministic node placement and the non- deterministic (stochastic) node placement. In this work we choose to place nodes in four different placements in order to cover all categories. A deterministic placement (Grid), a semi-deterministic (Biased Random) and two non- deterministic (Simple Diffusion and Random).

## 4.1. Deterministic Node Placement

In deterministic node placement, nodes are placed on exact, pre-defined points on a grid or in specific parts of the grid. Usually, deterministic or controlled node placement is specified by the type of nodes, the environment in which the nodes will deploy, and the application. Therefore, in applications like Sensor Indoor Surveillance Systems or Building Monitoring, nodes must be placed manually ([13]) (either by hand or by robots). In this work we employ Grid Placement as appears in Fig. 1. In this placement nodes are placed strictly on the lines of a grid.

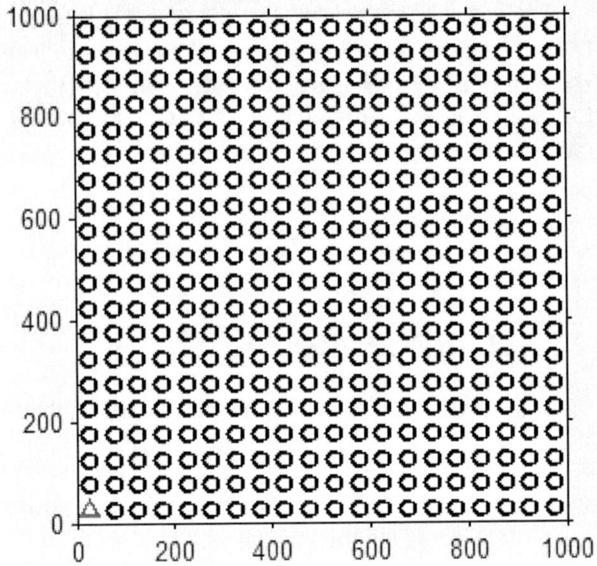


**Figure 1.** Grid Placement

## 4.2. Semi- Deterministic Node Placement

Semi- deterministic placement is the placement, where, although individual nodes are placed in a non- deterministic way on the grid (e.g random) the areas where nodes are going to be spread are deterministic. This means that in a microscopic way the placement of nodes is non- deterministic while in a macroscopic way the placement is deterministic. In this paper we employ biased- random placement, where nodes are placed in two specific areas (near source and near sink). Note that the actual node placement is performed in random way in these areas (Fig.2)

## 4.3. Non- Deterministic Node Placement

Deterministic placement is not so realistic when many sensor nodes are placed in a large area. In such a situation, stochastic placement is needed. In this paper we employ two stochastic placements. Simple Diffusion and Random placement

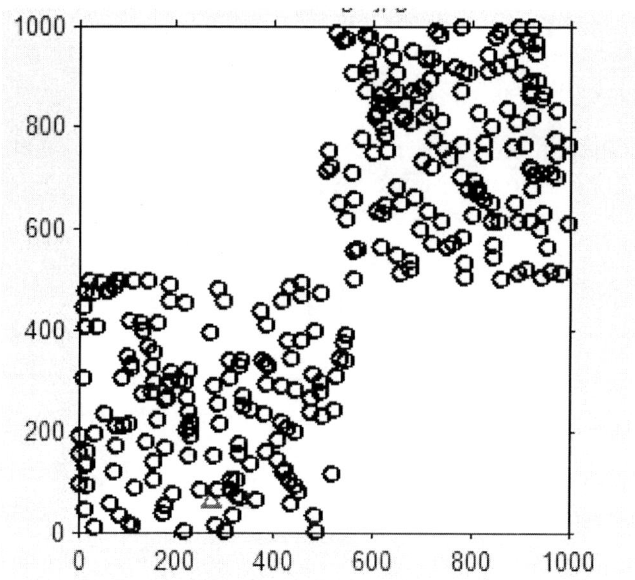


**Figure 2.** Biased Random Placement

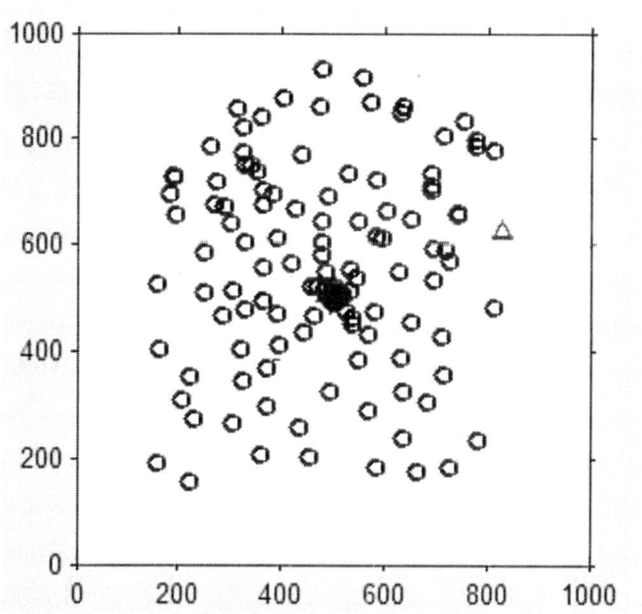


**Figure 3.** Simple Diffusion Placement

**Simple Diffusion**: This node placement emulates the distribution of nodes when they are scattered from air e.g from airplane (Fig. 3). Simple diffusion is analytically explained in ([15]).

**Random Placement**: This is a commonly used topology and sensor nodes are placed so that their density is uniform. (Fig. 4)

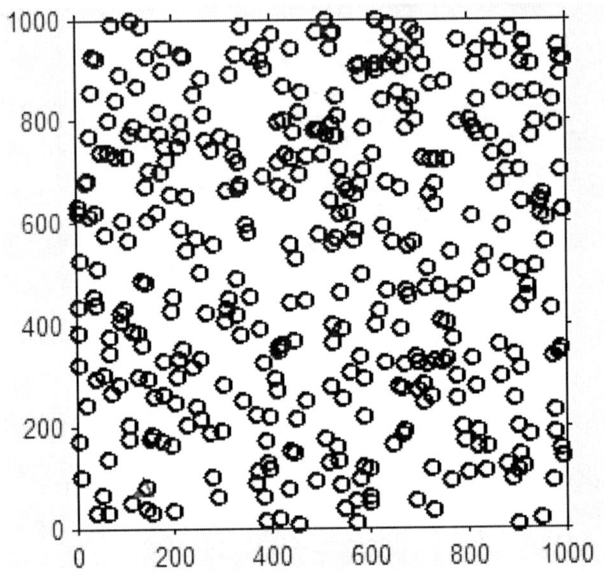


**Figure 4.** Random Placement

# 5. CONGESTION CONTROL ALGORITHMS

In different studies ([17], [19]) it is observed that the number of nodes with occupied queues grows, if congestion gets worse.

Several transport control schemes and algorithms have been proposed in the literature ([7], [8], [10],[17], [19], [20], [21], [22], [23], [24], [25]). Their objectives and approach differs. The vast majority of them ([7], [8] and [17] to [22]) react to congestion with rate limiting techniques ("traffic control" algorithms). Others deal with the problem by increasing the resources ([10], [26], [27]) ("resource control" algorithms). Some focus more on reliability issues instead of congestion

([7], [24], [25]) ("reliable data transport" algorithms). In our evaluation we consider one algorithm from each major category: ESRT ([7]) which is "traffic control" algorithm focusing on reliability, SenTCP ([8]) which is a "traffic control" algorithm focusing on congestion, and HTAP ([10]) which is "resource control" algorithm. We also employ Directed Diffusion ([9]) algorithm, an algorithm not explicitly designed for congestion control but it can be considered as so, since it employs a combination of "traffic control" and "resource control" techniques in order to provide multiple disjoint paths to the sink. Finally we also evaluate the performance of a generic routing algorithm, the "plain flooding".

Short description of employed algorithms follows.

## 5.1. Sen- Tcp (Sensor Based Tcp)

SenTCP is an open-loop hop-by-hop congestion control protocol with two special features:

- It jointly uses average local packet service time and average local packet inter-arrival time in order to estimate current local congestion degree in each intermediate sensor node. The use of packet arrival time and service time, not only precisely calculates congestion degree, but effectively helps to differentiate the reason of packet loss occurrence in wireless environments, since arrival time (or service time) may become small (or large) if congestion occurs.

- It uses hop-by-hop congestion control. In SenTCP, each intermediate sensor node will issue feedback signal backwards and hop-by-hop. The feedback signal, which carries local congestion degree and the buffer occupancy ratio, is used for the neighboring sensor nodes to adjust their sending rate in the transport layer. The use of hop-by-hop feedback control can remove congestion quickly and reduce packet dropping, which in turn conserves energy.

## 5.2. Esrt (Event To Sink Reliable Transport)

ESRT aims at providing reliability from sensors to sink while supporting congestion control simultaneously. It is an end-to-end algorithm trying to guarantee a desired reliability through regulation of sensors reporting frequency. It provides reliability for applications and not for each single packet. The sink uses congestion feedback from sensor nodes to broadcast a notification to adjust the reporting rate with two goals: i) to receive a sufficient number of nodes from the sink, and ii) to receive only as many packets as necessary in order to avoid congestion and save energy. ESRT runs on the sink, with minimal functionality required at resource constrained sensor nodes. ESRT protocol operation is determined by the current network state, based on the reliability achieved and congestion condition in the network. Firstly, it needs to periodically compute the factual reliability rr according to successfully received packets in a time interval. Secondly, ESRT deduces the required sensor report frequency fffrom rr. Finally, ESRT communicates ff to all sensors through an assumed channel with high power. ESRT identifies 5 characteristic regions:

- No Congestion, Low reliability
- No Congestion, High reliability
- Congestion, High Reliability
- Congestion, Low Reliability
- Optimal Operating Region (OOR) which essentially translates to No Congestion, Medium-High Reliability

The target is to identify network's current state and bring it in OOR (Optimal Operating Region). If the event-to-sink reliability is lower than the required, ESRT adjusts the reporting frequency of source nodes aggressively in order to reach the target reliability level as soon as possible. If the reliability is higher than required, then ESRT reduces the reporting frequency conservatively in order to conserve energy while still maintaining reliability. This self-configuring nature of ESRT makes it robust to random, dynamic topology of WSNs. An additional benefit resulted from ESRT is energy-conservation since it can control the sensor reporting frequency. ESRT presents some disadvantages:

- ESRT regulates report frequency of all sensors using the same value. It may be more reasonable to use different values, since each sensor node may have different contribution to congestion.
- It assumes and uses a channel (one-hop) with high power that will influence the on-going data transmission.

- It mainly considers reliability and energy-conservation. Feedback latency dependents on the network's size and may not scale in very large sensor networks.

## 5.3. Directed Diffusion

Directed Diffusion is a data centric protocol because all communication is for named data. All nodes in a directed diffusion-based network are application-aware. This enables diffusion to achieve energy savings by selecting empirically good paths (small delay) by caching and processing data in-network (e.g., data aggregation). Directed diffusion consists of four (4) basic elements:

- interests
- data messages
- gradients
- reinforcements

An interest message is a query from a sink node to the network, which indicates application demands. It carries a description of a sensing task that is supported by a sensor network. Data in sensor networks is the collected or processed information of an event (e.g. physical phenomenon), it is named (addressed) using attribute-value pairs, while a sensing task is diffused throughout the sensor network as an interest for named data. This dissemination sets up gradients within the network designed to "draw" events (i.e., data matching the interest). A gradient is direction state, created in each node that receives an interest. This direction is set toward the neighboring node from which the interest was received. Events start flowing towards the sinks of interests along multiple gradient paths. To improve performance and reliability, the empirically "good paths" (e.g small delay) are reinforced by the sink and their data rate increases. On the other hand unreliable paths (e.g high delay) are negatively reinforced and pruned off.

## 5.4. Htap (Hierarchical Tree Alternative Path)

HTAP is scalable and distributed framework for minimizing congestion and assuring reliable data transmissions in event based WSNs. As such it does not employ rate limiting actions, but tries to maintain a high level of packets data rate while minimizing packet losses. It is based on the creation of alternative paths from the source to sink ("resource control"), using the plethora of network's unused nodes, in order to safely transmit the observed data. The creation of alternative paths involves several nodes which are not in the initial shortest path from the source to the sink. The use of these nodes leads to a balanced energy consumption, avoiding the creation of "holes" in the network and prolonging network lifetime. The HTAP algorithm consists of four major parts.

- Flooding with level discovery functionality: Through this procedure, each node discovers its neighbor nodes and updates its neighbor table. In addition, sensor nodes are placed in levels from the source to the sink.

- Alternative Path Creation Algorithm: In order to avoid congestion, each candidate congested receiver sends a backpressure packet to the sender. Thus, the sender stops the transmission of packets to the candidate congested receiver and searches in its neighbor table to find the least congested receiver, in order to continue the transmission of data. The dynamic change of the receivers, leads to the creation of new routes from the source to the sink.

- The Hierarchical Tree Algorithm: A hierarchical tree is created, beginning at the source node. Connection is established between each transmitter and receiver using a 2-way handshake. Through this packet exchange, the congestion state of each receiver is communicated to the transmitter.

- Handling of Powerless (Dead Nodes): Special care is taken in HTAP algorithm concerning the nodes that their power is getting exhausted. Thus, when a node is going to exhaust its power, it is immediately extracted from the network and the tables of its neighbor nodes are updated.

# 6. PERFORMANCE EVALUATION

To evaluate the selected algorithms under the proposed topologies, a series of simulations using ns-2[28] simulator, has been conducted.

## 6.1. Simulation Environment

In all scenarios we choose to deploy nodes within a square area of size 1000m x 1000m. The results presented are the average of 20 runs for each measurement point. In each set of runs, the parameters of Table 1 were kept stable while increasing the number of nodes in the topology to make a dense network with strong connectivity.

**Table 1.** Simulation Parameters

| | |
|---|---|
| X distance (m) | 1000 |
| Y distance (m) | 1000 |
| Transmit Power (mW) | 600 |
| Receive Power (mW) | 600 |
| Sensitivity Threshold (dBm) | -81 |
| Path Loss Coefficient | 3.5 |
| Node CPU (MHz) | 4 |
| Radio Freq. (MHz) | 433 |
| Data packet | 128 bytes |
| Control Packet | 50 bytes |
| MAC layer | CSMA/CA |
| Initial Node Energy | 1 Joule |

The evaluation of node placements has been performed by studying four basic QoS parameters. These are: Average Packet Loss, Mean Packet's Delivery Delay, Average Data Rate, as well as the Percentage of Network's Remaining Energy after the point where the network, due to routing "holes" is not able to forward a single packets from source to sink.

## 6.2. Sentcp Evaluation

"Traffic control" is a method used in different algorithms to alleviate congestion in WSNs. SenTCP is one of these algorithms. In Fig. 5a we record the average data rate in all four examined topologies under SenTCP algorithm.

Initially, when 500 packets/s are injected in the overloaded network, the network experiences a heavy load situation and data rate falls rapidly in order to control the situation. SenTCP then slowly increases the rate until the occurrence of a new packet drop. It is clear that in all four topologies the network exhibits similar attitude and performance. This indicates that this parameter is slightly affected by nodes' placement. This is true, taking into account that SenTCP employs average local packet inter-arrival and packet service time, to detect congestion and "traffic contol" method to mitigate it.

Next we study packet drops. Packet drops is one of the most significant events in terms of performance control and their occurrence indicate a problem in the network. In Fig. 5b we record Average Packet Drops vs Simulation Time. According to this figure, the attitude of SenTCP algorithm is not affected by different placements. This result is an indication that "resource control" algorithms are not affected by different placements concerning their congestion mitigation ability.

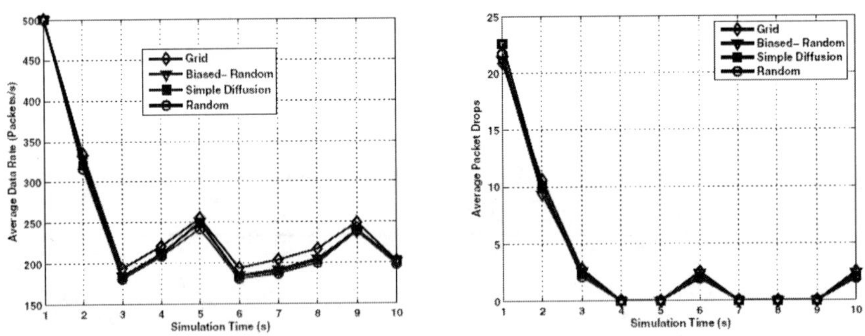


(a) SenTCP:Average Data Rate (packets/s)

(b) SenTCP:Average Packet Drops

**Figure 5.** SenTCP:Average Data Rate (packets/s) and Average Packet Drops

Also a significant parameter concerning performance control, is the minimization or reduction of delays in the network. In Fig. 6a we present the mean time for the transmission of packets from source to sink. It is obvious that as the hop number increases, mean time increases too, since hop count is bigger. Considering that, algorithms like SenTCP use the shortest path to transmit their data, it is expected that the placement which is able to provide the shortest path will exhibit the best performance. This happens in Simple Diffusion placement followed by Biased- Random.

The last parameter that we investigate is the percentage of network's remaining energy at the moment where the network is not able to transmit a single packet from source to sink (network stalls). This metric is an indication of whether the network has managed to uniformly utilize its resources, avoiding the creation of network connectivity "holes" (Fig. 6b).

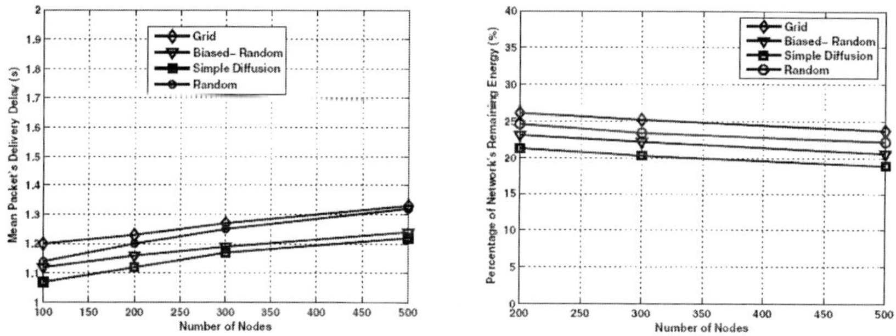


(a) SenTCP:Mean Packet's Delivery Delay (b) SenTCP:Percentage of Network's Remaining Energy

**Figure 6**. SenTCP:Mean Packet's Delivery Delay and Percentage of Network's Remaining Energy

As it is presented in this graph, the network utilizes most of its energy in Simple Diffusion placement while it consumes the least energy in Grid placement. This proves that in Simple Diffusion placement where the nodes are scattered around the sink, the network is able to utilize more uniformly its resources by finding more available paths from source to sink compared to Grid Placement.

## 6.3. Esrt Evaluation

ESRT is an algorithm that focuses on reliability. It is an end-to-end algorithm that runs on the sink and in case of congestion regulates report frequency (data rate) of all sensors using the same value. Fig. 7a presents the average data rate.

Since ESRT is an algorithm that throttles data rate in order to mitigate congestion, it is expected that average data rate is slightly affected by node placements. The same happens with packet drops (Fig.7b).

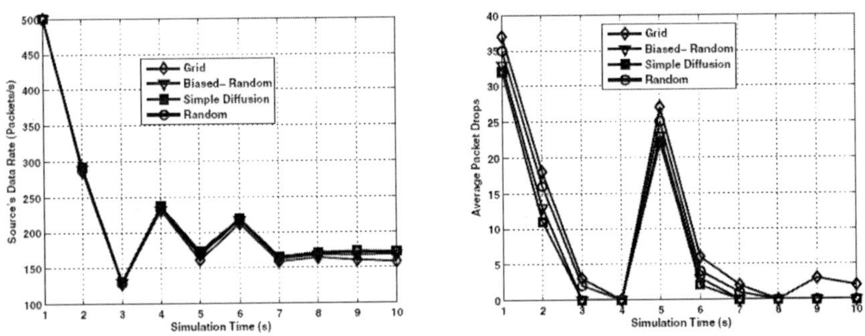


(a) ESRT: Average Data Rate (packets/s) (b) ESRT:Average Packet Drops

**Figure 7.** ESRT: Average Data Rate (packets/s) and Average Packet Drops

On the other hand mean packet's delivery delay" is a parameter that its attitude could be related with nodes placement. Efficient nodes placement can reduce the mean time for the transmission of packets. As it is presented in Fig. 8a, Simple Diffusion placement and Biased-Random placement (as with SenTCP) are the placements that provide the shortest paths from source to sink and normally present the least delay. Contrary, Grid Placement is a placement that provides longer paths and this is the reason that the delay in this placement is the biggest.

Concerning the percentage of network's remaining energy (Fig.8b) we notice that ESRT presents the same attitude as with SenTCP. ESRT when runs on Simple Diffusion and Biased- Random placements presents better performance in comparison with the other two placements. The reason is the same as with SenTCP. These placements provide a bigger number of disjoint paths from source to sink, that, when the nodes that form the initial paths are totally power exhausted, the network is still able to find other paths to forward data to sink.

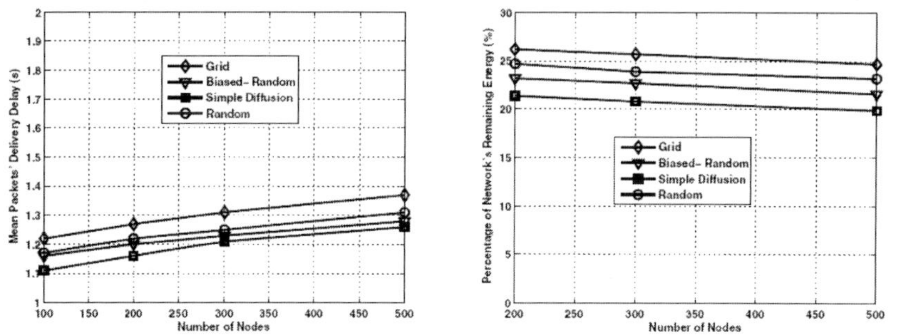


(a) ESRT:Mean Packet's Delivery Delay

(b) ESRT:Percentage of Network's Remaining Energy

**Figure 8.** ESRT:Mean Packet's Delivery Delay and Percentage of Network's Remaining Energy

## 6.4. Directed Diffusion Evaluation

Directed Diffusion is an algorithm that mitigates congestion in an indirect way. Initially, it sends an upstream data message to multiple nodes, forming multiple paths and then, with reinforcement and negative reinforcement attempts to reduce the number of paths, to a small number, based on their empirically observed performance. Through this reduction of paths, it controls the data rate of the paths and consequently the network's data rate.

Concerning the average data rate, Fig. 9a shows that it is affected by the topology. The Biased-Random placement scheme, exhibits the best performance followed by Simple Diffusion and Random Placement. The worst performance is exhibited by Grid Placement. The reason is that in Biased Random Placement, in a higher grade and Simple Diffusion in a lower, there is a high concentration of nodes one hop away from the sink. This leads to the creation of multiple disjoint "good" paths, which can be reinforced by the algorithm, to constantly maintain high data rates. On the other hand, in random placement as well as in grid placement, the nodes around the sink are limited. This leads to few high performance paths and the data rate is kept low.

Packet drops lead to higher latency paths which are not desirable especially in WSNs. Directed Diffusion uses negative reinforcement to prune off higher latency paths.

In Fig.9 b we record packet drops in all four topologies. It is clear that Simple Diffusion placement, due to the plethora of paths that constantly reinforces, presents null packet drops (after the initial injection of data packets in the network). On the other hand on the other topologies there are some packet drops, but negative reinforcement handles them quickly and efficiently.

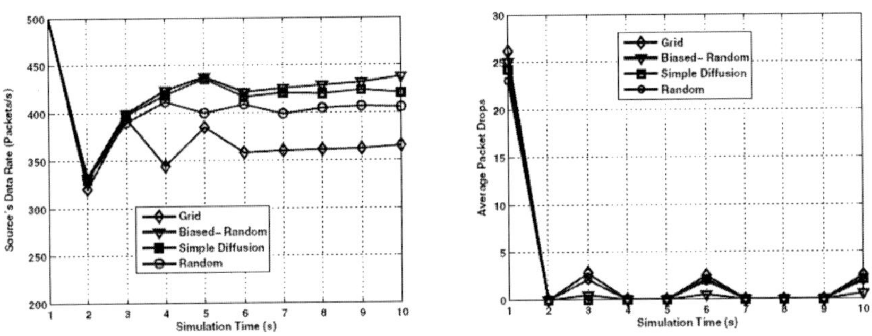


(a) Directed Diffusion: Average Data Rate (packets/s)

(b) Directed Diffusion:Average Packet Drops

**Figure 9.** Directed Diffusion: Average Data Rate (packets/s) and Average Packet Drops

Due to the nature of Directed Diffusion there is not much deviation between the four topologies, concerning mean packet's delivery delay. The reason is the reinforcement of high performance paths and negative reinforcement of low performance paths, which allows to the network to prune off high latency paths. In spite of this, Simple Diffusion followed by Biased- Random placement exhibits the best performance, in comparison with the other placements (Fig.10a), since it achieves nearly null number of packet drops (after the first injection of data packets in the network).

Directed diffusion presents much better performance compared to ESRT and SenTCP concerning the percentage of network's remaining energy. Comparing the performance of Directed Diffusion in different placements (Fig.10b) we record again, that when algorithm runs on placements like Simple Diffusion and Biased- Random, which are capable to provide many paths from source to sink, manages to utilize the network's resources uniformly. Good performance is also presented with random placement, since Directed Diffusion finds multiple paths for forwarding the data from source to sink. On the other hand, placement like Grid, are more vulnerable to the creation of network holes, due to the limited number of nodes near sources and sinks.

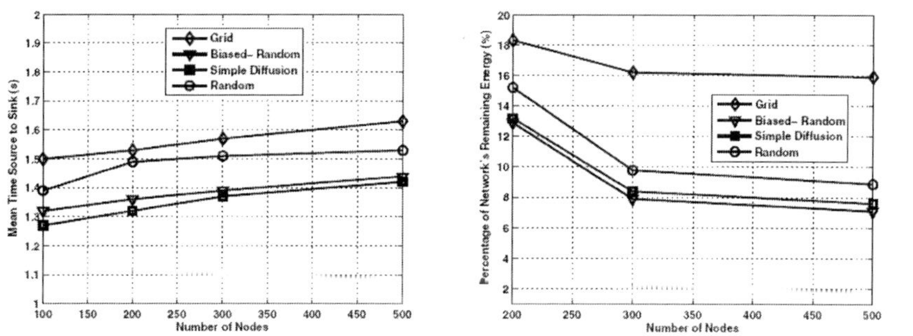


(a) Directed Diffusion:Mean Packet's Delivery Delay (b) Directed Diffusion:Percentage of Network's Remaining Energy

**Figure 10.** Directed Diffusion:Mean Packet's Delivery Delay and Percentage of Network's Remaining Energy

## 6.5. Htap Evaluation

By its nature, HTAP algorithm does not control the data rate of the sources, since it is an algorithm that mitigates congestion through the employment of sleep nodes ("resource control"). As it is expected, the network's data rate is kept stable in all simulation moments and for all topology schemes (Fig.11a).

Fig.11b presents HTAP's average packets drops. We observe that Biased- Random Placement exhibits the fewer packets drops compared to the other placements. The reason lies on the big number of nodes around sources and sinks. In this algorithm, in which the data rate is not reduced, the existence of many nodes one hop away from source and one hop away from sink is very important. Otherwise, if the nodes around sources and sinks are limited, the network will experience "hot-spot" congestion, at these nodes. This is what is happening in Random and Grid topology. Grid and Random placements face this situation very soon, since the number of nodes around the sources and sinks is limited, while Simple Diffusion faces this situation later due to the larger number of nodes (compared to Random and Grid topology) around the sink.

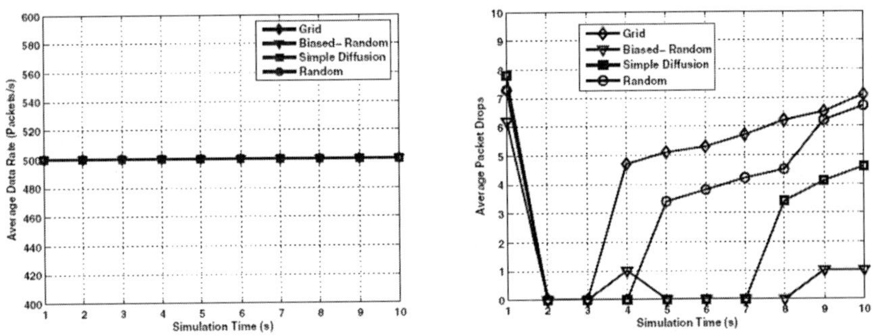


(a) HTAP: Average Data Rate (packets/s)

(b) HTAP:Average Packet Drops

**Figure 11.** HTAP:Average Data Rate (packets/s) and Average Packet Drops

In Fig.12a we observe that there is a deviation in mean packet's delivery delay between the four topologies, as the node density increases. Biased- Random and Simple Diffusion placements, seem to cope better with the increment of the number of nodes, as this increment creates more data paths. The reason in this case, as well, is the number of nodes around the sink, nodes that can directly forward the data to sink. On the other hand, as fewer nodes are around sink, latency increases due to the "hot spot" that appears near sink.

Concerning the percentage of network's remaining energy (Fig. 12b), we notice that HTAP exhibits the best performance compared to the other algorithms. The reason is the creation of alternative paths, that employs almost all nodes in the procedure of packets forwarding from source to sink. Comparing the performance of HTAP in the four placements we notice that beside Grid Placement, on the other three placements, HTAP algorithm exhibits very good performance and utilizes more than 90% of network resources. This is a strong indication that "resource control" algorithms can significantly increase the lifetime of a heavy loaded network.

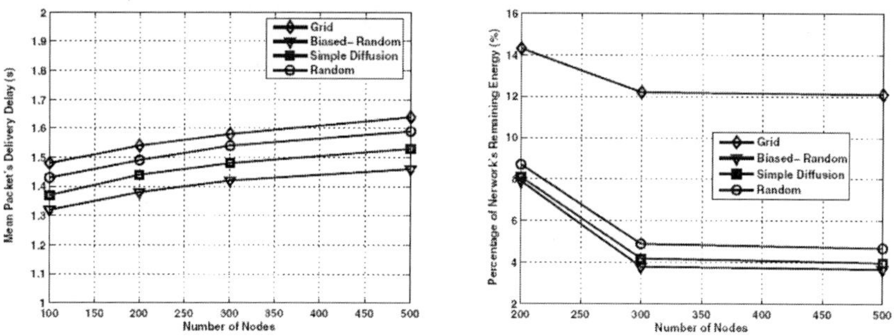


(a) HTAP:Mean Packet's Delivery Delay (b) HTAP:Percentage of Network's Remaining Energy

**Figure 12.** HTAP:Mean Packet's Delivery Delay and Percentage of Network's Remaining Energy

## 6.6. Flooding Evaluation

Flooding is generic routing algorithm. When flooding applies, each node forwards every message to every node that is in its radio range. Since it does not implement any "traffic control" functionality in case of congestion, then the sources data rate remains the same (Fig. 13a).

Fig. 13b presents average packet drops when flooding algorithm applies. As it is expected packets drops increase while time evolves. The reason lies on the functionality of flooding algorithm. That is the fact, that fills network with multiple copies of the same packet. Comparing just the placements, we notice that the placements that present the worst performance in the previous cases, when flooding applies present the best. Grid and Random are the placements with the fewer nodes around source. Counting that flooding algorithm, forwards every message to every node in its radio range, it is normal that the placements with the fewer paths, limit the number of packets in the network and thus the fewer drops appear.

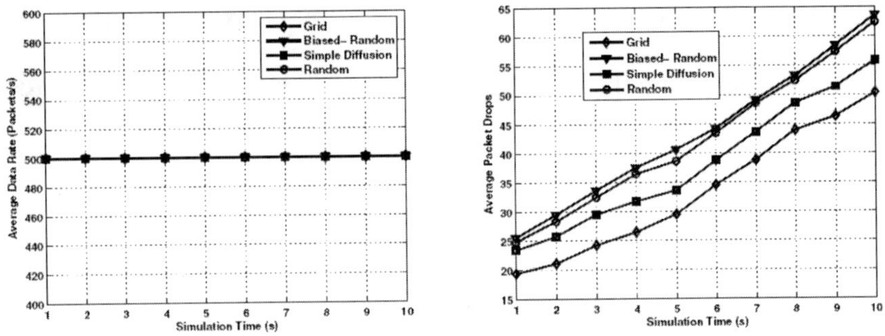


(a) Flooding: Average Data Rate (packets/s)

(b) Flooding: Average Packet Drops

**Figure 13.** Flooding:Average Data Rate (packets/s) and Average Packet Drops

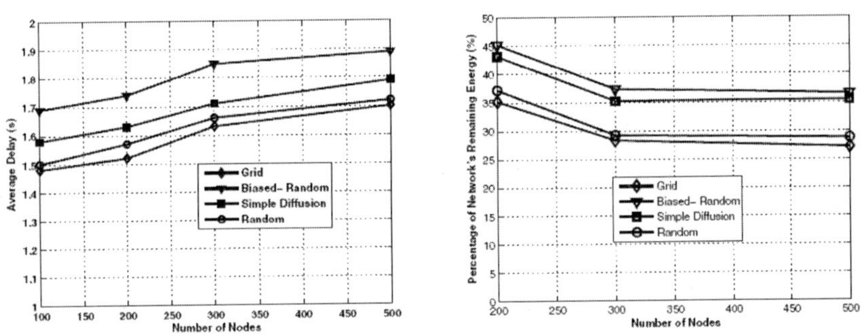


(a) Flooding:Mean Packet's Delivery Delay

(b) Flooding:Percentage of Network's Remaining Energy

**Figure 14.** Flooding:Mean Packet's Delivery Delay and Percentage of Network's Remaining Energy

The same attitude is depicted with mean packet's delivery delay (Fig. 14a). Again placements that create fewer paths are able to forward the data sooner.

Finally we check the percentage of network's remaining energy when the network stalls (Fig. 14b). In this case we also notice that placements with fewer nodes around source (Grid and Random placements) present better performance compared to the other placements. The reason lies on the operation of Flooding algorithm, (each node transmits each packets to all of its children). This leads to a bigger amount of transmissions from the nodes that have many children around source (this happen in Biased Random and Simple Diffusion placement) and soon these nodes are getting power exhausted. This leads to the creation of a "hole" around source and network "stalls".

# 7. CONCLUSIONS

In this paper we evaluated the performance of specific routing and congestion control algorithms when nodes are deployed under different placements. The algorithms we examined are ESRT ([7]), Sen- TCP ([8]), Directed Diffusion ([9]), HTAP ([10]) and Flooding, in Simple Diffusion, Random, Grid and Biased- Random Placements. Each algorithm represents a special category of congestion control and routing algorithm. ESRT is "reliable data transport" algorithm, SenTCP is a congestion control algorithm that mitigates congestion using "traffic control" method, Directed Diffusion discovers and maintains multiple high performance paths for transmitting packets from source to sink while HTAP is a congestion control algorithm that employs "resource control" method. Finally, flooding is a generic routing algorithm, that its functionality lies on the fact that each node tries to forward every message to every one of its neighbor nodes.

Simulation results show, that algorithms that employ multiple and alternative paths, either by default (Directed Diffusion) or as a response to a congestion situation (HTAP), for the transmission of data from source to sink are significantly favored by denser placements of nodes around source and sink since they can create many paths. This leads to fewer packet drops, while they extend significantly network's lifetime. On the other hand algorithms that always employ the shortest path for the transmission of packets from source are not affected by different node placements and in case of continuous heavy load they present shortest network's lifetime.

# 8. FUTURE WORK

Node placement is proven to be an effective way, for optimizing the performance in WSNs concerning "resource", congestion control algorithms. A future work on this subject would be the application of these placements on a real WSN environment and study the performance on a real network. Moreover, it would be worth studying what placements can assist "traffic control" algorithms to increase their performance. Initial results show that placements that create short paths from sources to sinks can assist in this direction. Finally, other parameters like fault tolerance, fault recovery, etc., are possible to be affected by different node placements. Examination of these issues constitutes part of our future work on the subject.

## ACKNOWLEDGMENT

This work has been partly conducted under the European Union Project GINSENG funded under the FP7 Program (FP7/2007-2013) grant agreement no 224282.

## REFERENCES

1. Alberto Cerpa, Jeremy Elson, Michael Hamilton, Jerry Zhao, Deborah Estrin, and Lewis Girod. Habitat monitoring: Application driver for wireless communications technology. In *SIGCOMM LA '01: Workshop on Data communication in Latin America and the Caribbean*, pages 20–41, New York, NY, USA, 2001. ACM.

2. E. Biagioni and K. Bridges. The applications of remote sensor technology to assist the recovery of rare and endangered species. *International Journal of High Performance Computing Applications, Special Issue on Distributed Sensor Networks*, April 2003.

3. Loren Schwiebert, Sandeep K. S. Gupta, and Jennifer Weinmann. Research challenges in wireless networks of biomedical sensors. In *MobiCom '01: Proceedings of the 7th annual international conference on Mobile computing and networking*, pages 151–165, New York, NY, USA, 2001. ACM Press.

4. H. T. Kung and D. Vlah. Efficient location tracking using sensor networks. In *IEEE Wireless Communications and Networking Conference (WCNC)*, pages 1954–1962, March 2003.

5. R.R. Brooks, P. Ramanathan, and A. M. Sayeed. Distributed target classification and tracking in sensor networks. *Proceedings of the IEEE*, 91(8):1163–1171, 2003.

6. B.C. Srean, J.S. Silva, L. Wolf, R. Eiras, T. Voigt, U. Roedig, V. Vassiliou, and G. Hackenbroich. Performance control in wireless sensor networks: the ginseng project - [global communications news letter]. *Communications Magazine, IEEE*, 47(8):1 –4, august 2009.

7. Y. Sankarasubramaniam, O. Akan, and I. Akyildiz. ESRT: Event-to-Sink Reliable Transport in Wireless Sensor Networks. In *MobiHoc '03: Proceedings of the 4th ACM International Symposium on Mobile Ad Hoc Networking & Computing*, pages 177–188, New York, NY, USA, 2003. ACM.

8. C. Wang, K. Sohraby, and B. Li. SenTCP: A Hop-by-Hop Congestion Control Protocol for Wireless Sensor Networks. *IEEE INFOCOM (Poster Paper)*, March 2005.

9. Chalermek Intanagonwiwat, Ramesh Govindan, Deborah Estrin, John Heidemann, and Fabio Silva. Directed Diffusion for Wireless Sensor Networking. *IEEE/ACM Transactions on Networking*, 11(1):2–16, 2003.

10. Charalambos Sergiou, Vasos Vassiliou, and Andreas Pitsillides. Reliable Data Transmission in Event-Based Sensor Networks During Overload Situation. In *WICON '07: Proceedings of the 3rd International Conference on Wireless Internet*, pages 1–8, Austin, Texas, October 2007.

11. V. Vassiliou and C. Sergiou. Performance Study of Node Placement for Congestion Control in Wireless Sensor Networks. In *New Technologies, Mobility and Security (NTMS), 2009 3rd International Conference on*, pages 1 –8, dec. 2009.

12. Charalambos Sergiou and Vasos Vassiliou. Energy utilization of HTAP under specific node placements in Wireless Sensor Networks. In *Wireless Conference (EW), 2010 European*, pages 482 –487, 12-15 2010.

13. Mohamed Younis and Kemal Akkaya. Strategies and Techniques for Node Placement in Wireless Sensor Networks: A Survey. *Ad Hoc Networks*, 6(4):621–655, 2008.

14. Stavros Toumpis and Leandros Tassiulas. Optimal Deployment of Large Wireless Sensor Networks. *IEEE Transactions on Information Theory*, 52(7):2935–2953, 2006.

15. Mika Ishizuka and Masaki Aida. Performance Study of Node Placement in Sensor Networks. *ICDCSW '04: Proceedings of the 24th International Conference on Distributed Computing Systems Workshopsd*, pages 598–603, March 2004.

16. H. Zhang and J. C. Hou. Is deterministic deployment worse than random deployment for wireless sensor networks? In *INFOCOM 2006. 25th IEEE International Conference on Computer Communications. Proceedings*, pages 1–13, April 2006.

17. A. Woo and D. E. Culler. A Transmission Control Scheme for Media Access in Sensor Networks. In *Proceedings of the 7th annual International Conference on Mobile Computing and Networking (MobiCom'01)*, pages 221–235, New York, NY, USA, 2001. ACM Press.

18. Sameer Tilak, Nael B. Abu-Ghazaleh, and Wendi Heinzelman. Infrastructure tradeoffs for sensor networks. In *WSNA '02: Proceedings of the 1st ACM international workshop on Wireless sensor networks and applications*, pages 49–58, New York, NY, USA, 2002. ACM.

19. Chenyang Lu, Brian M. Blum, Tarek F. Abdelzaher, John A. Stankovic, and Tian He. RAP: A Real-Time Communication Architecture for Large-Scale Wireless Sensor Networks. Technical report, Charlottesville, VA, USA, 2002.

20. Cheng Tien Ee and Ruzena Bajcsy. Congestion Control and Fairness for Many-to-One Routing in Sensor Networks. In *SenSys '04: Proceedings of the 2nd international conference on Embedded networked sensor systems*, pages 148–161, New York, NY, USA, 2004. ACM.

21. Bret Hull, Kyle Jamieson, and Hari Balakrishnan. Mitigating Congestion in Wireless Sensor Networks. In *SenSys '04: Proceedings of the 2nd International Conference on Embedded Networked Sensor Systems*, pages 134–147, New York, NY, USA, 2004. ACM.

22. Chieh-Yih Wan, Shane B. Eisenman, and Andrew T. Campbell. CODA: Congestion Detection and Avoidance in Sensor Networks. In *SenSys*

*'03: Proceedings of the 1st international Conference on Embedded Networked Sensor Systems*, pages 266–279, New York, NY, USA, 2003. ACM Press.

23. Karthikeyan Sundaresan, Vaidyanathan Anantharaman, Hung-Yun Hsieh, and Raghupathy Sivakumar. ATP: A Reliable Transport Protocol for Ad Hoc Networks. *IEEE Transactions on Mobile Computing*, 4:588–603, 2005.

24. Fred Stann and John Heidemann. RMST: Reliable Data Transport in Sensor Networks. In *Proceedings of the First International Workshop on Sensor Net Protocols and Applications*, pages 102–112, Anchorage, Alaska, USA, April 2003. IEEE.

25. Chieh-Yih Wan, Andrew T. Campbell, and Lakshman Krishnamurthy. P.S.F.Q: A Reliable Transport Protocol for Wireless Sensor Networks. In*WSNA '02: Proceedings of the 1st ACM international workshop on Wireless Sensor Networks and Applications*, pages 1–11, New York, NY, USA, 2002. ACM Press.

26. Jaewon Kang, Yanyong Zhang, and Badri Nath. TARA: Topology-Aware Resource Adaptation to Alleviate Congestion in Sensor Networks. *IEEE Transactions on Parallel and Distributed Systems*, 18(7):919–931, 2007.

27. C. Sergiou and V. Vassiliou. DAlPaS: A Performance Aware Congestion Control Algorithm in Wireless Sensor Networks. In *Telecommunications (ICT), 2011 18th International Conference on*, pages 167 –173, may 2011.

28. The Network Simulator ns-2 (v2.1b8a). http://www.isi.edu/nsnam/ns/, October 2001.

2

# POWER OPTIMIZATION FOR WIRELESS SENSOR NETWORKS

**A.R. Pinto[1], L. Bolzani Poehls[2], F. Vargas[2] and C. Montez[3]**

[1] Universidade Estadual Paulista, UNESP, Brazil

[2] Engineering School, Catholic University of Rio Grande do Sul, PUCRS, Brazil

[3] PGEAS, Universidade Federal de Santa Catarina, UFSC, Brazil

## 1. INTRODUCTION

Wireless Sensor Network (WSN) is a denomination that covers a lot of variations in compositions and deployment. A typical sensor network consists of a large number of low cost, low power distributed devices, called nodes, deployed in the environment being sensed and controlled (Stankovic et al., 2003). In other words, this kind of network is composed of a huge number of tiny nodes able to communicate with each other that can be used to monitor hazardous and inaccessible areas. Thus, each node consists of processor, memory, wireless antenna, battery and the sensor itself. Nodes can sense scalars from the environment such as temperature, acoustic and light, but may also process and transmit them by radio. The network can be classified as homogeneous or heterogeneous, which would mean that some specific nodes present special hardware or software configuration, but even in homogeneous networks, to collect, store and process data from the WSN's nodes, a special node, called Base Station (BS), is necessary.

Most of the currently adopted technologies for WSNs are based on low-cost processors, resulting in limited energy budget and restricted memory space. In many applications, it is expected that the sensor node last for a long time because in most of the cases these networks are used in remote areas and recharging and/or replacing power supply units is considered difficult or prohibitive due to hazardous and inaccessible places where they are supposed to operate. Further, due to the availability of cheap hardware and various possibilities for the radio communication frequency, numerous topologies for WSN can be adopted (Akyildiz, 2002; Ilyas & Mahgoub, 2005; Oliver & Fohler, 2010).

As previously mentioned, the nodes in these networks are usually inexpensive and therefore WSNs may be composed of a huge number of sensor nodes, which themselves are deployed inside and/or around the phenomenon that one desires to monitor. Not necessarily the sensor node's geographic position is previously known in all adoptions, because, when the nodes need to work in hazardous or inaccessible areas, it might be impossible to avoid their random deployment. Thus, if this type of application is adopted, the use protocols and approaches that can self-organize and self-optimize energy consumption of a large number of nodes that cooperate in order to achieve a global goal, becomes necessary (Akyildiz, 2002). Summarizing, WSN are desired to present the following characteristics (Ilyas & Mahgoub, 2005):

- Self-organization;
- Short range wireless communication and multi-hop routing;
- Large number of nodes and cooperative efforts between WSN nodes;
- Different WSN topologies, which frequently change due to battery depletion and node faults;
- Constrained resources such as energy budget, processing and memory.

The characteristics above and the capacity of interaction with the environment distinguish WSNs from other ad-hoc wireless networks. WSNs, due to the scarce software and hardware resources, are application oriented; thus, WSN applications are developed focusing on a specific problem's solution. The collaborative efforts between nodes are necessary to the correctly use the WSN's resources, this effort can also increase the WSN's lifetime (Stankovic et al., 2003; Hadim & Mohamed, 2006).

It is important to highlight that the strategy of deploying a large number of unreliable nodes presents advantages when compared with deploying few expensive but very reliable nodes (Ilyas & Mahgoub, 2005). These advantages are listed below:

- Larger spatial resolution;
- Higher fault tolerant degree achieved throughout distributed techniques;
- More uniform coverage;
- Ease of deployment;
- Reduced energy consumption;
- Increased network lifetime.

WSNs present a high degree of environmental interaction, depending on where sensor nodes are deployed, implicit and explicit temporal restrictions apply. In this context, data freshness is an important concept that dictates how long a sensed scalar can be considered useful and when it can be discarded. In the following example, the information gathered by a security application based on WSN technology, identifies any person who enters into a certain area of the building in a certain time, all data that exceeds this time limit, is not useful. Despite having time constraints, due to the high node density, non-determinism, noise and constrained WSN resources, it becomes extremely difficult to guarantee real-time properties (Stankovic et al., 2003). Therefore, these special constraints impose that no hard deadlines are considered for WSN application. Consequently, critical real-time systems are out of scope for this kind of network (Koubaa et al., 2009; Oliver & Fohler, 2010).

WSN can be considered an innovative paradigm, which permits the emergence of several new monitoring applications, but introduces challenges intrinsic to this technology. Some of these challenges are (Akyildiz, 2002; Stankovic et al., 2003; Molla & Ahamed, 2006; Yick et al., 2008;Ilyas & Mahgoub, 2005):

- *Paradigm change:* WSN are basically deployed in order to collect scalars from the environment and support control applications. The WSN application must sense the environment and, sometimes, act in one way or the other on the environment. Thus, it is considered critical to obtain a cooperative behaviour of thousands of sensor nodes where the data from just one node may not be important. The sensor nodes do not have a permanent identification address, due to the fact that generally messages are not sent to a specific node but to a space or area (based on the message's content). Users can be interested in the information about a specific monitoring area, thus the sensed data from a specific node may not be important, representing the data centric approach in WSNs. The need of physical environment interaction also implies important differences between WSNs and ordinary ad-hoc networks, thus classical distributed system techniques are not applicable to WSNs. Real-time requirements, noise, high fault occurrence and non determinism also impose a new group of approaches that must deal with these constraints (Molla & Ahamed, 2006).

- *Resource Constraints:* As noted above, WSNs face severe resource constraints. The main resource constraints are: limited energy budget, restricted CPU clock, restricted memory as well as network bandwidth. These characteristics impose the application of new solutions. The fact that WSN topologies are composed of a huge number of nodes represents a new issue that had not been considered in simple ad-hoc networks. For instance, trade-off approaches that aim at guaranteeing energetic economy and real-time characteristics became necessary (Yick et al., 2008).

- *Unpredictability:* There are many uncertainties that may affect a WSN. Firstly, WSN are deployed in environments with multiple uncontrollable events. Secondly, wireless communication is sensitive to noise; data lost due to radio interference and several physical errors is common to networks employed in harsh environments. Thirdly, nodes are not individually dependable. Further, it is not always possible to properly calibrate the nodes before employment; routing structures such as paths and the connectivity can be dynamically added or excluded during the WSN time of functioning. The addition or removal of nodes might be necessary due to permanent faults or battery depletion. Additionally, the energy level in some node can significantly vary even during the initial deployment. Last, the nodes might be physically removed due to environment causes or intentional controlling, thus a network restructuring must would be necessary.

- *Self-*:* One of the biggest challenges is to create the WSN's vision in the network application layer. Due to the fact that WSNs are deployed to operate with few or none human intervention, self-* characteristics like self-organization, self-optimization and self-healing become necessary (Huebscher & McCann, 2008; Oliver &

Fohler, 2010). These characteristics are easily listed as challenge, however are extremely difficult to achieve.

- *High scale/density:* There are several WSN approaches that consider a large number of nodes in order to overcome hardware or software faults, thus there is a minimum number of nodes that is necessary to guarantee the WSN's service. The main challenges include: the processing of this large number of generated data, the assurance that the particular WSN requires the minimum desirable density, and the development of solutions that require the lowest density and energy consumption in order to maximize the WSN's lifetime. A WSN based on a large number of nodes that are deployed in large areas is considered a large-scale system. Due to its characteristics, these systems are subject to faults, noise, which sometimes can be caused by the WSN itself, and other uncertainties. Moreover, when a WSN is deployed, it might be self-operational and present self-maintenance, due to the fact that human intervention is sometimes very expensive or even impossible. Therefore, all these characteristics impose several conflicting goals. These challenges can be increased by the technology scaling, where the industry's minimization tendency (Akyildiz et al., 2008).

- *Real-time:* WSNs operate in the real world, thus real-time features are necessary to guarantee the correct functionality. These systems present implicit real-time constraints. The response time of its tasks is also important, thus the system tasks must be finished as fast as possible. Several WSNs present explicit real-time constraints. For example, a structural monitoring application imposes explicit deadlines for the data sensing (Kim et al., 2007). However, due to the large number of nodes, non-determinism and noise, it might be extremely hard to guarantee real-time properties.

- *Security:* WSNs can be used in safety critical applications, thus their security is an essential issue to be considered. Denial of Service techniques can be easily executed over a WSN. Moreover, coordination and real time communication approaches do not consider security issues. Thus, some intruder can easily exploit these WSN security faults. The great dilemma is how to implement security techniques that need large computational resources in a technology that deals with severe hardware constraints.

In this scenario, where nodes are likely to operate on limited resources, power conservation is considered one of the most important concerns of these networks and different strategies and protocols need are adopted in order to deal with it (Gholamzadeh & Nabovati, 2008). In more detail, network lifetime can be enhanced if the system's software, including different layers and protocols, is designed in a way that lowers the consumption of energy (Gholamzadeh & Nabovati, 2008). Several techniques have been proposed in literature in order to decrease the power consumption of WSNs. These techniques are related to different aspects of sensor networks, from hardware platform to Medium Access Control (MAC) protocol, routing and topology control.

This Chapter is structured as follows: Section 2 summarizes the main applications where WSNs are deployed and their hardware characteristics. In Section 3, the main MAC layer approaches proposed in literature are described. Section 4 presents the routing strategies proposed in order to provide power optimization and consequently increasing WSN's lifetime, while Section 5 introduces Transmission Power Control approaches. Section 6 introduces Autonomic approaches and finally, in Section 6 the final remarks on the optimization of WSNs are presented.

## 2. WSN APPLICATIONS AND HARDWARE CHARACTERISTICS

WSNs are considered an application oriented technology, thus approaches that are developed for some specific application usually cannot be used for different uses. Important points related to the hardware characteristics of the nodes must be considered in order to guarantee the suitable node for a specific application. In more detail, aspects related to the type of processing unit as well as communication, power supply and sensing devices must be considered, when the nodes for a specific application are defined.

Considering aspects related to the processing unit, usually a microcontroller or microprocessor is adopted. In order to choose the ideal microcontroller for the system and due to the fact that microcontrollers with high performance imply higher power consumption, the designer must consider the desired performance level. Another important point is associated to the fact that microcontrollers usually support different operational modes, such as active, idle and sleep mode, which directly affect the power consumption of the node. There is also an attractive design option that suggests splitting the workload between two low power microcontrollers in such way that one of the microcontrollers is responsible for the sensing control, while the other performs the networking tasks related to controlling the RF interface and running the algorithms (Chou & Park, 2005). Finally, techniques like the one called Dynamic Volta Scaling (DVS) can be adopted (Karl & Willig, 2005). DVS dynamically adapts the microcontroller's power supply voltage and operating frequency to meet the processing requirement, thus trading off performance and power supply for energy savings.

Different communication devices using mediums like radio frequency or optical communications, for example, can be adopted to exchange data between nodes. For communication, both a transceiver and a receiver are required for the sensor nodes. The essential task of these devices is to convert a bit stream coming from a microcontroller into radio waves and vice versa. In more detail, the transceiver is normally regarded the largest power consumer and optimizing its power can result in significant improvement for the system as a whole (Chou & Park, 2005). There are several factors that affect the power consumption characteristics of a transceiver, including its type of modulation scheme, data rate, transmission power and the operational duty cycle (Gholamzadeh & Nabovati, 2008). Many transceivers allow the user to set the power level. In general, transceivers can operate in the following distinct modes of operation: Transmit, Receive, Idle and Sleep; allowing switching between them and consequently realizing energy savings. Note that the switching between the operating modes has to be managed, taking into account the fact that waking up a transceiver from the Sleep mode and making it go to Transmit mode requires some start-up time and start-up energy. Thus, switching a node into Sleep mode is only beneficial, if the energy necessary for the node to comeback into an active mode is smaller than the energy saved during the Sleep mode, which implies that the time to the next event is sufficiently large.

Regarding the power supply device, a battery is used in most of the cases, playing a vital role in determining the sensor node's lifetime. Thus, one of the most important factors that a designer must consider is the rate capacity effect, which is related to the discharge rate or the amount of current draw from the battery. Drawing higher current than the rated value leads to a significant reduction in battery life, due to the fact that the diffusion of electrolyte falls behind the rate at which they are consumed at the electrodes. It is important to highlight that most of the applications of WSNs involve deploying sensor nodes in harsh and remote environment and therefore it is difficult to use ordinary recharging schemes for batteries. In particular cases, an alternative is to adopt external energy resources like sunlight or wind.

Finally, sensors in WSNs translate physical phenomena to electrical signals and can be classified as analog or digital devices depending on the type of output they produce. Basically, there are several sources of power consumption in a sensor: signal sampling and conversion of physical signals to electrical ones, signal conditioning, and analog to digital conversion (Gholamzadeh & Nabovati, 2008). Passive sensors, such as temperature sensors, consume less power than active sensors, like sonar, which need energy to send out a signal to probe the observed object. Indeed, the sampling rate is important and higher frequency sampling requires more energy. In this context, sensors should acquire a measurement sample only if needed, when needed, where needed and with the right level of fidelity (Raghunathan et al., 2006). This strategy reduces the energy consumed in the subsystem and sometimes reduces the processing and communication load as well. Thus, the use of mechanisms able to change the bit resolution of measurement samples and the sampling rate as well as using adaptive spatiotemporal sampling, exploiting redundancy and correlations models to predict a measurement instead of actually making it and finally, hierarchical sensing, can provide power consumption optimization.

In the next paragraphs, several WSN applications will be briefly presented. According to the temporal requirements of the applications, they may present significant differences in the applied algorithmic solutions. For instance, an environment monitoring application can require deadlines of minutes, while in a military application temporal validity is much smaller. Some kinds of applications need periodic sensing and sending, while other applications need an event driven approach.

Possible applications of WSNs include environment monitoring, military, domotic and industrial monitoring and control. For instance, an application of habitat monitoring is presented in (Mainwaring et al., 2002). The presented WSN has been deployed on the Great Duck Island. Its main goal is to correlate the measurements of some microclimate data (temperature, light and humidity) with the bird nest activity on the island. This application presents relaxed real-time requirements, thus the main goal of the Great Duck Island application is the maximization of the network's lifetime, as it is expected that the WSN's infrastructure stays active during months or even years without human intervention.

Therefore, the intervals between sending messages and between one sensing and another may be reduced significantly.

Another example for a WSN's application is structural monitoring, in this case a linear WSN topology was used to monitor the Golden Gate Bridge's structure, and thus a routing technique had to be applied to assure the messages delivery to the BS at one end of the construction. This application is based on accelerometer sensors that detect modifications in the physical structure of the bridge (Kim, S. et al., 2007).

Finally, other kinds of WSNs' utilization are stated below:

- *Automotive industry:* Cars are equipped with sensor and actor networks, which can interact with highway or street WSN infrastructures in order to increase traffic efficiency or automate toll payment;

- *Monitoring and automation of factory systems:* Industrial robots can be equipped with thousands of wired sensors. These sensors must be connected to a central computer. The high economic cost and the mobility restrictions of wired sensor are favour the utilization of WSNs in this kind of robots.

- *Intelligent housing:* WSNs permit that houses can be equipped with movement, light and temperature sensors, microphones used for voice activation and pressure sensors in chairs are also examples of WSN utilization in building automation. Thus, air temperature, natural and artificial lighting and other components can be tuned according to specific user needs;

- *Merchandise tracking:* Logistic and transportation companies may use WSN technology in order to track ships, transporters, containers or single goods that are being transported;

- *Precision agriculture:* Irrigation control and precise pesticide application are possible with the help of WSN utilization on farmlands;

- *Harsh area monitoring:* Exploration and monitoring of harsh areas may be possible throughout the use of WSNs;

- *Freshwater quality monitoring:* WSNs may be used for freshwater monitoring due to their non-intrusiveness and small size;

- *Military application:* Position and movement control of troops and vehicles, target detection, non-human combat-area monitoring as well as landmine removal or building exploration are just some examples of possible utilizations of WSNs for military applications.

To conclude, WSNs can be applied in different types of applications and the selection of the suitable hardware depends on the systems' requirements, the available resources and the environment where the network should operate.

# 3. MAC LAYER APPROACHES

As previously mentioned, the lifetime maximization of WSNs is one of the most important concerns when dealing with the use of WSNs. This is mainly related to the fact that sensor nodes are considered unavailable when the battery level is depleted. In this context, it is important to note that communication among nodes is the major energy consumer process in WSNs. A significant portion of the node's energy is spent on radio transmissions and on listening to the medium for anticipated packet reception (Gholamzadeh & Nabovati, 2008). In other words, sending or receiving messages requires significantly more energy than data processing or the acquisition by the sensors. Moreover, a single medium for communication is shared by the nodes and network performance largely depends on how efficiently and fairly these nodes share the medium. MAC protocol controls the communication nodes in WSNs and regulates access to the shared wireless medium such that the performance requirements of the underlying applications are satisfied (Sohraby et al., 2007). Thus, a careful definition of protocols and algorithms for efficient communication has become one of the most important issues in WSNs in order to improve their lifetime. Basically, the MAC protocol must be energy efficient and must try to reduce the following issues related to energy consumption phenomena (Demirkol, 2006):

- *Packet collision*: When one node receives more than one packet at the same instant, it is considered that a packet collision occurred. Therefore, all packets must be discarded and transmitted again.

- *Overhearing*: When some node receives packets that are addressed to another sensor node overhearing occurred;

- *Control packet overhead*: The use of control packets in order to coordinate the WSN must be minimized;

- *Idle listening*: Idle listening occurs when some node is in the listening mode of a channel that is not being used.

- *Over-emitting*: Over-emitting is caused when the message delivery fails due to the destination node's inactivity.

It is important to highlight that the collisions of messages is considered the most critical aspect, which causes the discarding of all involved messages and forces the network to retransmit increasing its energy consumption. Thus, an energy-efficient MAC protocol must avoid collisions and reduces the energy dissipation related to idle channel sensing, overhearing and overhead to a minimum (Ilyas & Mahgoub, 2005).

Regarding the types of communication patterns, four different types can be identified for WSNs (Demirkol, 2006):

- *Broadcast*: Generally BSs use the broadcast communication pattern (sink) to transmit certain information to all nodes under its controls. The broadcast information must contain consults, software upgrades or some control packet. The broadcast pattern can only be used, when all destination nodes are inside the radio coverage of the transmitter node;

- *Local gossip*: Local gossip is to be considered when nodes sense some event and send it to other nodes that are located inside the same location (same cluster). This kind of communication occurs when one node sends its messages to its neighbours, inside the same coverage area;

- *Convergecast*: This type of pattern is used when a group of sensors sends their packets to a specific node. The destination node can be a cluster-head, a fusion centre or a BS;

- *Multicast*: Some scenarios imply that messages need to be sent to a group of sensor nodes, thus just the sensors that belong to this group receive the message.

In the next subsection the IEEE 802.15.4 Standard and the ZigBee technology are described and the main MAC approaches able to reduce power consumption are summarized as well. Another subsection, will present further MAC approaches aiming at the optimization of MSNs.

## 3.1. Ieee 802.15.4 Standard and The Zigbee Technology

The main goal of the ZigBee technology is to enable WSNs composed of large number of nodes to function with reduced energy consumption. Most WSN technologies like *Mica Motes* use ZigBee in order to achieve higher lifetime levels for their WSN applications.

The ZigBee network architecture is based on the *Open Systems Interconnection* (OSI), however exclusively the more important layers were implemented. ZigBee adopts the IEEE 802.15.4 standard, which only defines the lower layers: the physical layer and the MAC layer (ZigBee, 2010).

The physical layer may operate on two frequencies 868/915MHz or 2.4GHz, with 16 channels and 250Kbps of maximum transmission rate. The IEEE 802.15.4 MAC layer is based on the Carrier Sense Multiple Access with Collision Avoidance (CSMA-CA) mechanism. Note that ZigBee technology differs from other wireless technologies due to several reasons: lower data transmission rate, lower energy consumption, lower cost, higher self organization and more flexible network topologies (ZigBee, 2010).

The IEEE 802.15.4 standard has been proposed in 2003 and has become a *de facto* standard for low energy consumption and low data rate transmission networks. The IEEE 802.15.4 MAC protocol supports two kinds of operational modes that can be selected by a central node called Personal Area Network (PAN) coordinator. The two modes are:

- *Beaconless mode*: where MAC protocol functions are based on a CSMA/CA without beacon packet;
- *Beacon mode*: where beacons are periodically sent by the PAN coordinator in order to synchronize nodes that are associated with it and to delimit a *superframe. D*uring the *superframe*duration all node transmission must occur. Moreover, during the contention period of this frame the MAC protocol is ruled by the slotted CSMA/CA. The IEEE 802.15.4 with beacon mode can use the

synchronization and the contention free period that is based on a guaranteed slot time.

Thus, the ZigBee Alliance is responsible for the ZigBee technology standardization. In more detail, the application and network layers are defined by the ZigBee Alliance itself, while the physical and MAC layers are based on the IEEE 802.15.4 standard. ZigBee may also consider time synchronization according to an optional *superframe* structure; the ZigBee technology possesses an address scheme that can handle up to 65.000 nodes. Moreover, three kinds of topologies are supported: *star*, *mesh* and*cluster tree*. The *Star* topology is considered the simplest topology and is based on a many-to-one communication topology, which means that all nodes are covered by the PAN coordinator antenna, and are able to send the messages in just one hop. However, the *mesh* and *cluster tree* topologies rely on a routing protocol in order to deliver the messages to the PAN coordinator. *Mesh* topology does not allow cluster-heads and nodes to communicate with each other. Still different, the *cluster tree* topology is based on the organization of nodes into clusters. Basically, the *star* topology is considered simpler than the *mesh* and *cluster-tree* ones due to the fact that no routing protocol is necessary (ZigBee, 2010).

IEEE 802.15.4 standard is based on CSMA/CA MAC algorithm, its beaconless mode does not impose the sending of a periodical beacon by the PAN coordinator (IEEE 802.15.4, 2008).

Two parameters are considered in the beaconless mode: the first is denominated NB that is the Number of times CSMA/CA is required to backoff, the second is called Backoff Exponent (BE), standing for the number of backoff periods a device must wait until it can access the communication channel.

The first step of the CSMA/CA algorithm is the initialization of NB and BE. After the initialization, the MAC layer must wait a random period of 0 to ($2^{BE}$ - 1) and then require the Clear Channel Assessment (CCA) from the physical layer. When the channel is considered occupied by other device, the NB and BE is incremented by 1 by the MAC layer, though the MAC algorithm must guarantee that BE never grows above *macMaxBE*. Moreover, when NB´s value is above *macMaxCSMABackoffs*, the CSMA/CA algorithm must quit and returns the access channel failure status.

Finally, the Beaconless CSMA/CA algorithm is sensitive to three parameters: *macMaxBe* (standard value 5), *macMaxCSMABackoffs* (standard value 4) and *macMinBE* (standard value 3). These standard values for the parameters may help to decrease energy consumption due to the fact that devices try to send just five times before the transmission's abortion, however incrementing these values tends to increase the network's communication efficiency. Thus, the IEEE 802.15.4 protocol is not able to deal with dense network topologies, them being networks that are based on a large number of nodes.

## 3.2. Other Mac Approaches

Several other MAC approaches have been proposed in literature in order to provide the reduction of power consumption in WSNs. In the next paragraphs the main solutions that explore the optimization of MAC protocols are summarized.

Techniques used in the MAC layer of WSNs often involve the use of Time Division Multiple Access (TDMA) and Duty Cycles (DC).

The main idea behind the TDMA technique is to divide the time spent by devices over channel accesses into so called time slots, each one used exclusively by one device. Therefore, by applying this technique, every device, before sending any messages, needs to book such a slot of time in advance. A TDMA MAC protocol is proposed in (Shi & Fapojuwo, 2010). This technique is based on a cross-layer optimization involving MAC and physical layers. The main goal of the presented technique is to reduce the overall energy consumption based on a TDMA scheduling with the shortest frame length for clustered WSNs.

In order to also reduce energy consumption, the TDMA MAC protocol is used in (Wu, Y. et al., 2010). Here, the main focus is to schedule the sensor nodes with consecutive time slots at different radio states, such as: transmitting, receiving, listening, sleeping, and idle. Due to the fact that sensor nodes consume different levels of energy at each state the optimum scheduling of these states could achieve the reduction of energy consumption.

FlexiTP is a TDMA MAC protocol that schedules node messages based on the so-called sleep scheduling approach (Lee et al., 2008). The sleeping scheduling scheme requires that sensor nodes to exclusively transmit and receive packets at their own time slot and turn into sleep state until their slot's turn is up again. FlexiTP also provides routing, time synchronization tasks and sensor nodes may sense as well as route data.

PEDAMACS is another TDMA MAC protocol designed for multihop WSNs (Ergen and Varaiya, 2006). It can improve the network's lifetime by several years when compared to other MAC protocols such as random access protocols that may reach months or just days of network lifetime. However, this TDMA protocol does not present a good performance when applied to WSNs with dynamic topologies, as they are common in harsh environments.

Complementarily, the DC technique divides the operating time of devices in two periods: active and inactive, also denominated sleeping time. The shorter the period of activity is in comparison with the period of inactivity, the longer the devices remains inactive and consequently achieves greater energy savings. As downside, a consequent reduction of the maximum transmission rate in the network is to be observed. If, on the one hand, TDMA enables devices to become more organized in order to avoid collisions; on the other hand, the DC technique is able to avoid cases where a node becomes simultaneously active with other nodes that had been inactive before, preventing a node to wait for messages that will never arrive, and finally avoiding the waste of energy.

These techniques generally allow the protocols to deal with collision, idle listening and over-emitting problems, but have overheads associated to sending and processing of control messages. Such extra-costs can be unnecessary in applications where the density of the network is small and where few devices transmit simultaneously. In this scenario, contention-based protocols such as Carrier Sense Multiple Access with Collision Avoidance (CSMA-CA) seem to be more suitable. Note that the CSMA/CA MAC protocol present in the IEEE 802.15.4 standard (IEEE 802.15.4, 2008) is inefficient for large networks.

A Rotational Listening Strategy (RLS) for Wireless Body Network (WBN) is presented in (Tseng et al., 2011). WBNs are a special kind of WSNs that are deployed over a human body area in order to sense and transmit scalars as for example the body's temperature. The RLS approach is based on the division of channel access partitioning it into mini-slots that are allocated to nodes.

Another type of WBN application is presented in (Omeni et al., 2008), where a MAC protocol implemented in hardware by a 0.13μm CMOS process is described. In order to avoid collisions with nearby transmitters in a wireless body network, a standard listen-before-transmit technique is being used. The time slot overlap handling was reduced based on a wakeup-fallback approach.

A different Wireless Body Networks (WBN) MAC protocol is proposed in (Otal et al., 2009). The main goal of this MAC protocol is the maximization of the battery lifetime of each individual body sensors while maintaining the reliability and message latency of data transmissions at the same level. To do so this MAC protocol is based on a cross-layer fuzzy rule scheduling algorithm and a energy-aware radio activation policy for realistic medical applications.

An Energy-Aware Hybrid Data Aggregation (EDHDAM) technique is presented in (Kim, M. et al., 2011). It aims at minimizing the energy problem in asynchronous MAC-based WSNs. The nodes closer to the sink spend more energy than other, this is due to the fact that they receive and send more data to the WSN's sink then nodes that are far away from the sink. Thus, the EDHDAM technique is designed to adaptively control the number of data transmissions in order to avoid the before mentioned downside.

A game-theoretic MAC approach for WSNs is presented in (Zhao et al., 2009). The MAC of nodes in this technique is based on an incompletely cooperative game mode. This approach denominated G-MAC, where time is divided into super-frames, each super-frame having two parts: an active part and a sleeping part. During the sleeping part, all nodes turn-off their radios to save energy and during the active part, if some node has packets to send, these will pass on the channel that is based on the incompletely cooperative game.

Multiple cross-layer protocols that integrate MAC and WSN's network layers are presented in (Rossi & Zorzi). All these MAC protocols are cost-aware regarding residual energies, link conditions, and queue state. The routing layer chooses the best relay candidates based on the MAC protocol information. In this manner, the number of in-range devices, that compete for one channel as well as the interference are reduced.

A technique named S-MAC, a medium access control based on coordinated adaptive sleep scheduling, is presented in (Ye at al, 2004). S-MAC tries to avoid the overhearing problem by low-duty-cycle operations in a multi-hop WSN. The S-MAC approach organizes the sensor nodes into virtual clusters based on common sleep scheduling in order to reduce the control overhead and enable traffic-adaptive wake-ups.

Finally, MRMAC is a MAC protocol that reduces the end-to-end delay as well as the energy consumption in WSNs. The approach presented in (Hong *et al.,* 2010) reduces the end-to-end delay based on two metrics: Next Packet Arrival Time (NPAT) and Medium Reservation Information (MRI). When a sender transmits a packet, the NPAT and MRI metrics are enclosed in the packet in order to make possible that its intended receiver reserves the medium. The simulations presented by show that the MRMAC approach is able to significantly reduce, both end-to-end delay and energy consumption.

# 4. ROUTING APPROACHES

WSNs can be adopted for a wide range of applications, but in all of them, the main task of the nodes is to sense and collect data, process it and transmit the information to the site where it is possible to analyse the monitored parameters. To efficiently achieve this task, it is required the development of an energy efficient routing protocol to set up paths between the nodes and the data sink (Sohraby et al., 2007). Due to the fact that sensor nodes are energy constrained, great part of the WSN's protocols aim at minimizing the energy required for communications. Basically, the environment characteristics coupled with scarce resources and energy limitation make the routing problem very challenging. Thus, the path selection must be such that the lifetime of the network is maximized. In this context, different strategies can be adopted in order to face with this problem. One simple strategy is to avoid bad-quality routes because unreliability of wireless links has an adverse effect on their performance. Link failures and packet losses lead to many retransmissions and therefore, result in higher power consumption.

A clustering protocol called REACA is presented in (Quan et al., 2007). REACA's functioning is divided into two cycles: the first cycle is dedicated to the network configuration while the second cycle handles the message transmission. The cluster-head to be chosen is based on the battery level of all nodes that compose the cluster. Thus, the node that presents the highest battery energy level is chosen to function as cluster-head. Moreover, a routing algorithm is proposed and REACA is validated by mathematical analysis.

EARQ by (Heo et al., 2009) is a routing protocol based on the WSN's energy level. EARQ is able to guarantee dependability, temporal constraints and energy economy. The main goal of EARQ is to use the path with the greatest energy level inside a WSN. Its authors prove by simulations that EARQ may be implemented in a WSN for industrial application. However, EARQ was not validated for any WSN prototype. Another cluster-tree WSN routing protocol was proposed by (Alippi et al., 2009).

The approach called MMSPEED is a routing protocol that is able to guarantee probabilistic Quality of Service (QoS) metrics in WSNs. It was proposed by (Felemban et al., 2006), considers different options of speed delivery in the time domain and guarantees package delivery. Several dependability requirements are provided, which are based on several path options. The end-to-end requirements are provided in a located fashion; this is desirable in terms of scalability and adaptability in dynamic and dense WSNs. However, the utilization of geographical routing imposes that nodes need to know their geographical localization. Thus, the proposing authors considered that each node would possess GPS devices or distributed localization algorithms. This results in considerable problems, as GPS devices do not work properly in indoor environments and distributed localization algorithms impose an extra overhead due to the extra package exchange, since the nodes need to periodically broadcast their geographical localization.

The q-Switch, a simple path routing algorithm proposed in (Wu, X. et al., 2008), is a routing technique used to support the non-uniform node distribution strategy that is used to mitigate the energy hole problem in WSNs. Its authors also show that in a circular multi-hop WSN with non-uniform node distribution and constant data sending the unbalanced energy consumption is unavoidable.

An approach denominated the Energy Efficient Broadcast Problem (EEBP) in ad hoc wireless networks is presented in (Li et al., 2004). The EEBP's idea can be described by the following phrase: in a given an ad hoc wireless network, find a broadcast tree such that the energy cost of the broadcast tree is minimized. Its authors consider that all the network´s nodes present a fixed level of transmission power. As solution three routing approaches aiming at the minimization of the network´s consumption are proposed.

A sleep scheduling solution called Green Wave Sleep Scheduling (GWSS), which has been inspired by synchronized traffic lights, is presented in (Guha et al., 2011). The main goal of this approach is to support the WSN´s routing duty cycling. A green wave is a moving sequence of consecutive active states (green lights), and some packet may move in a sequence of active nodes. Thus, nodes in sleep mode are compared to red lights, and packages may not be routed through a sleeping node. Its authors show that, considering large WSNs arranged in structured topologies, GWSS achieves almost the same end-to-end latency as that of non-sleep-scheduling WSNs.

# 5. TRANSMISSION POWER CONTROL APPROACHES

Power conservation is so important because nodes are usually operating on limited batteries. As previously mentioned, MAC protocols are able to manage energy consumption during WSN communication, which is the most energy-consuming event in WSNs. However, one interesting solution in order to increase WSN's lifetime is based on adjusting its nodes' transmission power. On the one hand, maintaining the lowest possible transmission power represents a interesting solution in order to minimize the energy consumption and consequently increase the network's lifetime. On the other hand, the lowest possible transmission power can increase the WSN's vulnerability to the interference fluctuations caused by bad Signal-to-Interference-plus-Noise-Ratio (SINR) (Kim & Know, 2008). Extensive empirical studies confirm that the radio communication's quality between low power sensor devices varies significantly with time and environment. This phenomenon indicates that the existing topology control solutions, which use static transmission power, transmission range and link quality, might not be effective in the physical world (Lin et al., 2006). In this context, online transmission power control techniques that take into account environment variations have become essential in order to address this issue.

Several Transmission Power Control (TPC) approaches have been proposed in the literature. Basically, the TPC algorithm can reduce the energy consumption and improve the channel capacity. In more detail, TPC solutions use a single transmission power for the whole network, not making full use of the configurable transmission power provided by radio hardware to reduce energy consumption or assume that each node chooses a single transmission power for all the neighbours, which is know as neighbour-level solutions. Indeed, most existing WSNs use a network-level transmission power for each node.

There are many TPC studies, which mainly focus on improving the channel capacity (Monks et al., 2001; Ho & Liew, 2007). Recently, experimental studies (Don et al., 2004) (Lin et al., 2006) have shown that TPC reduces energy consumption in low-power WSNs. In Power Control Algorithm with Backlisting (PCBL), each node sends packets at different transmission power levels to determine the optimal transmission power based on the Packet Reception Ration (PRR). In (Lin et al., 2006), an Adaptive Transmission Power Control (ATPC) algorithm is proposed in order to achieve the optimal transmission power consumption for specified link qualities. Employing a ATPC algorithm, the Received Signal Strength (RSS) and the Link Quality Indicator (LQI) for radio channels are used to estimate the optimal transmission power level, and employ a feedback-based ATPC algorithm to dynamically adjust the transmission power over time. Thus, the result of applying this algorithm is that every node knows the proper transmission power level to use for each of its neighbours and every node maintains good link qualities with its neighbours by dynamically adjusting the transmission power through on-demand feedback packets.

However, the effect produced by different inference sources must be considered when the goal is the implementation of WSNs in the physical world. Many WSN devices available on the marked operate on the 2.4GHz ISM band and are vulnerable to the interferences from other wireless networks such as the IEEE 802.11 WLANs or the IEEE 802.15.1 Bluetooth (Kim & Know, 2008). Generally, the transmission power of the WSN devices is lower than that of WLAN or Bluetooth devices. Therefore, the TPC algorithm for WSNs has to carefully consider the interferences caused by other 2.4GHz wireless devices, which can cause significant performance degradation. In this context, a practical TPC algorithm for WSNs, named Interference Aware Transmission Power Control (I-TPC) algorithm has been proposed in (Kim & Know, 2008). The I-TPC algorithm is based on the idea that each node adjusts the RSS target to provide the acceptable SINR when interferences are detected. In more detail, the I-TPC algorithm consists of two functional procedures: the two-tier transmission power control and the RSS target adjustment. Initially, the proper RSS target, which may satisfy the desired PRR is determined. Based on this RSS target, each node tries to adjust its transmission power to keep the RSS value within the upper and the lower RSS target values by using the two-tier transmission power control procedure. The net effect of this operation is that the proposed algorithm tries to reach a satisfying link quality quickly even if there are small-scale link quality variations. When the interference is detected, the RSS target and the transmission power are increased immediately by the RSS target adjustment procedure to provide an appropriate SINR.

Two different local algorithms to individually adjust the nodes' transmission power are presented in (Kubish et al., 2003). Such local approaches do not require any particular MAC protocol or dedicated protocol for route discovery. The so-called Local Mean Algorithm (LMA) implements that each node periodically sends a *life message* and all receiving nodes respond with *life acknowledge messages.*Before sending new information each node counts the received acknowledge messages and compares this number to the value set as thresholds. In the case the node received less messages then the inferior threshold, the transmission power is increased by factor $A_{inc}$ for every node missing to achieve the threshold. If this number is in the range between the minimum and maximum threshold no changes to transmission power are made. Similarly, the Local Mean Number of Neighbours (LMN) algorithm works with life and life acknowledge messages. In addition to the LMA approach the *life acknowledge message* contains the nodes own count of neighbours. Thereby each node receives a number of messages containing a value indicating the numbers of neighbours of the sending node, then calculates a mean value from all the received messages and uses this value as well as the number of nodes that responded to its *life message* to calculate the so called *NodeResp* value to be compared with the thresholds and, if the case, the transmission power is adapted as described in the LMA technique. These two techniques are compared to fixed and global algorithms and, in the given indoor scenario, are outperforming the fixed approaches while reaching only about half the lifetime of networks employing global algorithms such as the Equal Transmission Power (ETP) algorithm. It is noted that comparing such approaches to ETP the local algorithms are almost competitive when looking at the network's confidence level and on top are scalable and easily implementable, which global algorithms are not.

Finally, a Transmission Power Self Optimization (TPSO) technique is presented in (Lavratti et al., 2012). It basically consists of an algorithm able to guarantee the connectivity as well as an equally high Quality of Service (QoS) concentrating on the WSN's Efficiency (*Ef*), while optimizing the necessary transmission for data communication in each node. The technique aims at adjusting each node to use the lowest possible transmission power while maintaining the connectivity to the WSN and the reliability of the network as a whole. This trade-off between the WSN's *Ef* and the data transmission energy consumption is evaluated in different EMI environments. Its decentralized algorithm runs on the application layer and uses an *Ef* value calculated, which adopts the number of received messages and the estimate of sent messages. This *Ef* is compared with the targeted *Ef* in order to decide about adjustments to the node's transmission power. Experimental results show that the automatic adaption presents advantages when compared with approaches using fixed transmission power. It is shown that the technique is able to guarantee the trade-off between *Ef* and power consumption. The TPSO behaviour is shown in Figure 1. It is possible to notice that the energy dissipated by a node with fixed transmission power is much higher than the energy consumed by a node running the TPSO algorithm. The session values in the x-axis represent the elapsed time.

Figure 2 shows the WSN's *Ef* and the WSN's energy consumption with respect to the WSN using the five pre-defined transmission power levels and to the WSN adopting the TPSO technique.

We can observe the effectiveness of the TPSO technique with respect to the use of pre-fixed transmission power levels. In detail, we can see this network using the maximum transmission power level reaches about 80% of *Ef*, but consumes about 50mW.s, while the WSN adopting the TPSO algorithm reaches the same *Ef* consuming only about 25 mW.s.

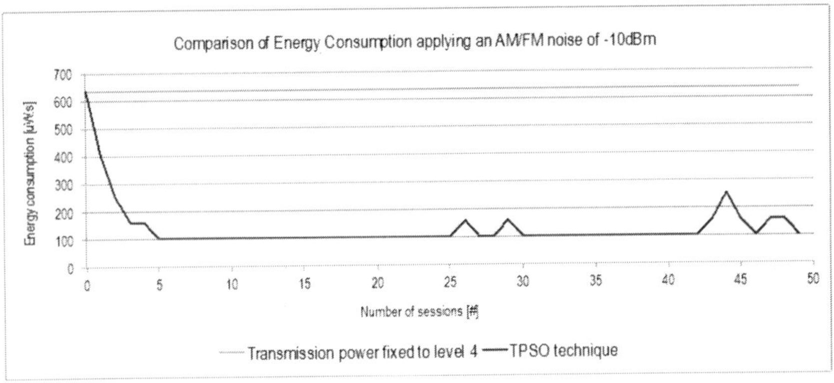


**Figure 1.** Comparison of the energy consumption of the TPSO technique with respect to WSN operating with the transmission power fixed to level 4 (**Lavratti et al., 2012**).

Figure 3 depicts the *Ef* of two WSNs, one with the transmission power level set to 0 and one with the TPSO technique. The WSN with the fixed transmission level is able to reach an average *Ef* of 46.4%, while the other network is achieving 86.6%. As the WSN with the TPSO technique is switching to higher transmission power levels to cope with the introduced noise it needs 253% more energy to reach the higher level of effectiveness.

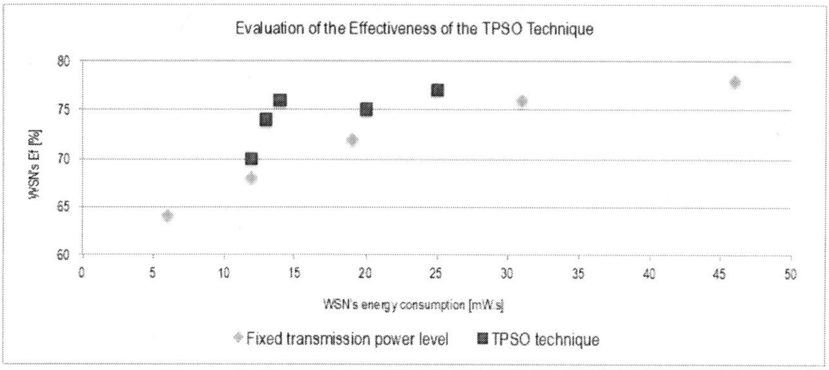


**Figure 2.** Evaluation of the TPSO technique with respect to WSN's Ef and energy consumption applying AM/FM noise of -14 dBm (**Lavratti et al., 2012**).

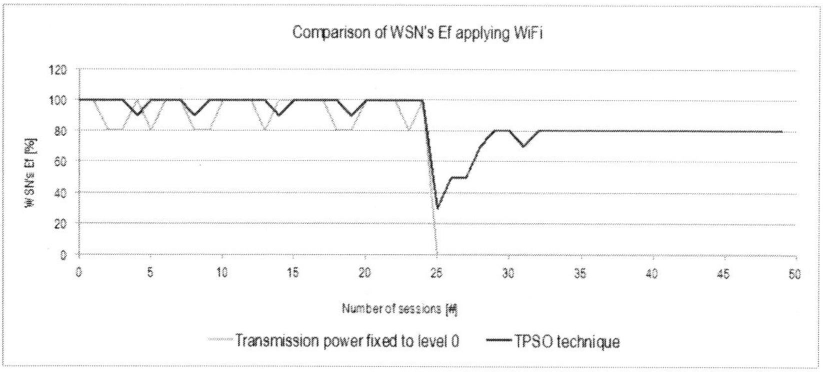


**Figure 3.** Comparison of the WSN's Ef of the TPSO technique with respect to WSN operating with the transmission power fixed to level 0 (**Lavratti et al., 2012**).

The results obtained during the experiments demonstrated the convenience of using the self-optimization algorithm instead of setting the maximum transmission power level. When a WSN without the TPSO technique is considered, the transmission power is set at the beginning of the communication and remains the same during its entire lifetime. This characteristic can be negative considering a WSN in a real environment where the inherent noise is not necessarily constant. Therefore, due to the fact that the inherent environment noise is completely variable and random, the TPSO technique will always guarantee the lowest possible transmission power during the communication and the target_*Ef* when it is possible (Lavratti et al., 2012).

# 6. AUTONOMIC APPROACHES

IBM introduced the term autonomic computing in 2001 to describe computer systems able to self-manage themselves (Kephart and Chess, 2003). The main properties of approaches categorized as "self-*" are:

- self-configuring
- self-optimizing
- self-healing
- self-protection.

Each one of them is described in (Huebscher and McCann, 2008). A brief definition is presented below:

- Self-configuration: the system's ability to configure itself according to high level goals;
- Self-optimization: the system can decide to start a change in the system as pro-active, in order to optimize the performance or quality of service;
- Self-healing: the system detects and diagnoses problems, which can be either faulty bits in a memory chip or a software error;

- Self-protection: the system is able to protect itself against malicious attacks or unauthorized changes.

Even though dense WSNs present several advantages, self-management characteristics are required in order to deal with the management of a large number of nodes. Self-management techniques are part of autonomic-computing methodologies, which can also be used to manage WSNs with conflicting targets (energy efficiency, self-organizing, time constraints and fault tolerance). The main goal of self-management is the development of a computing system that does not need the human intervention to operate. This way, computing systems are able to self-organize and self-optimize themselves, once they follow global objective dictated by a system administrator (Pinto e Montez, 2010).

For instance, in dense WSNs composed of several sensor nodes in a star network topology, in case the network presents conflicting goals (increase dependability and energy efficiency, while meeting time constraints), the conventional IEEE 802.15.4 protocol does not seem to be able to deal with the complexities. For example, when the number of nodes in a network is increased, in order to achieve better reliability the WPAN may be congested, and fewer messages arrive in the base station on time. In order to demonstrate the WSNs behaviour in this situation, experiments using TrueTime simulator [1] - were performed. Two metrics called *Ef (efficiency)* and *QoF* have been adopted. While efficiency is a metric that measures the ratio between sent and received messages; *QoF* represents the average number of received messages by the base station over a certain timespan. Figure 4 shows that when density network is increased, QoF increases slowly, but communication efficiency quickly decreases.

(Pinto e Montez, 2010) propose a Genetic Machine Learning Algorithm (GMLA) aimed at applications that make use of trade-offs between different metrics. The main goal of the GMLA approach is to improve the communication efficiency, in a communication environment where the network topology is unknown to the base station.

Simulations were performed on random star topologies assuming different levels of faults. ObservingFigure 5 it can be notice that the communication efficiency maintains the same level when IEEE 802.15.4 is used. However, when GMLA is used, it is possible to notice a gain of almost 10% in communication efficiency.

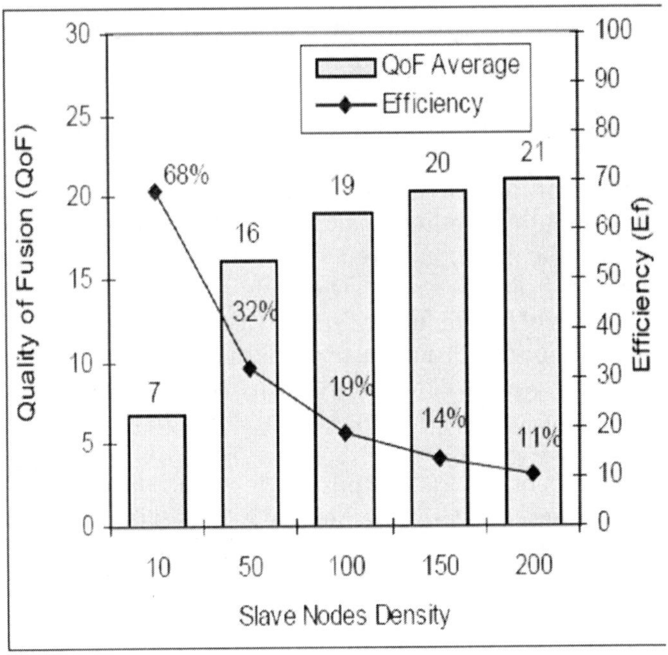


**Figure 4.** IEEE 802.15.4 simulated behaviour.

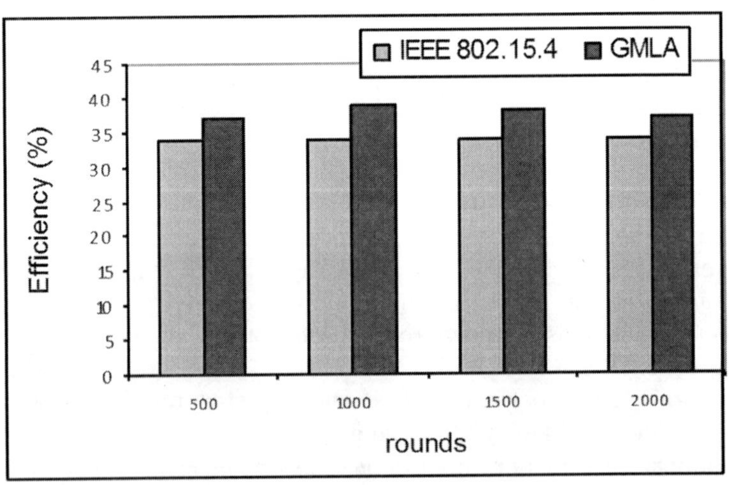


**Figure 5.** Comparison of GMLA and IEEE 802.15.4.

It is possible to notice that IEEE 802.15.4 presents a static behavior, and that it does not learn better communication patterns when topology changes are faced (Figure 6).

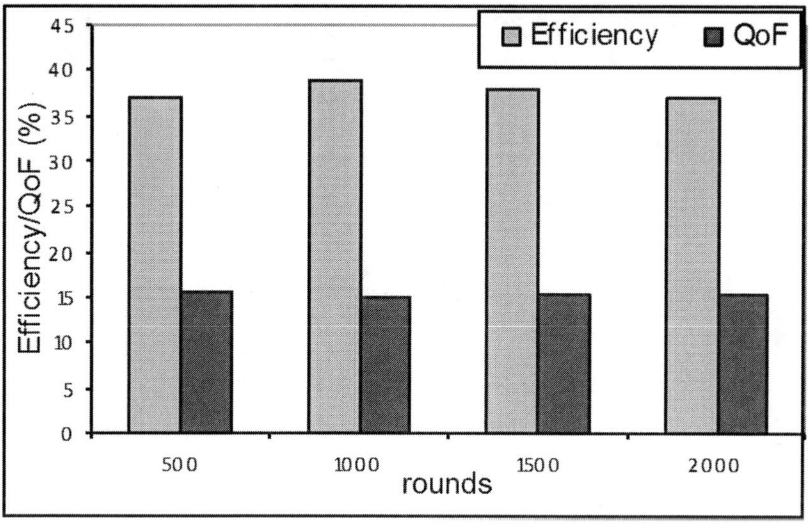


**Figure 6.** GMLA Efficiency and QoF values.

An analysis of Figure 7 indicates that the *QoF* is maintained at almost the same level, in all simulations. However, the higher level of *Ef* was achieved with 1,000 round simulations. This may be explained through GMLA´s learning behaviour, which tries different configurations when longer simulations are run.

Consequently, it is possible to conclude that the GMLA approach is able to do the trade-off between*QoF* and *Ef* and GMLA uses lower levels of energy than IEEE 802.15.4. However, this approach is only suitable for applications with a homogeneous signal throughout the entire monitoring area.

Also in (Pinto e Montez, 2010), a Variable Offset Algorithm (VOA), which targets the optimization of the communication efficiency in dense WSNs with star topology, is proposed. The VOA can be easily implemented into IEEE 802.15.4 devices, as it is a light middleware that is implemented at the application layer. The main target of VOA is the communication efficiency through the use of random offsets before the transmission of data by the slave nodes.

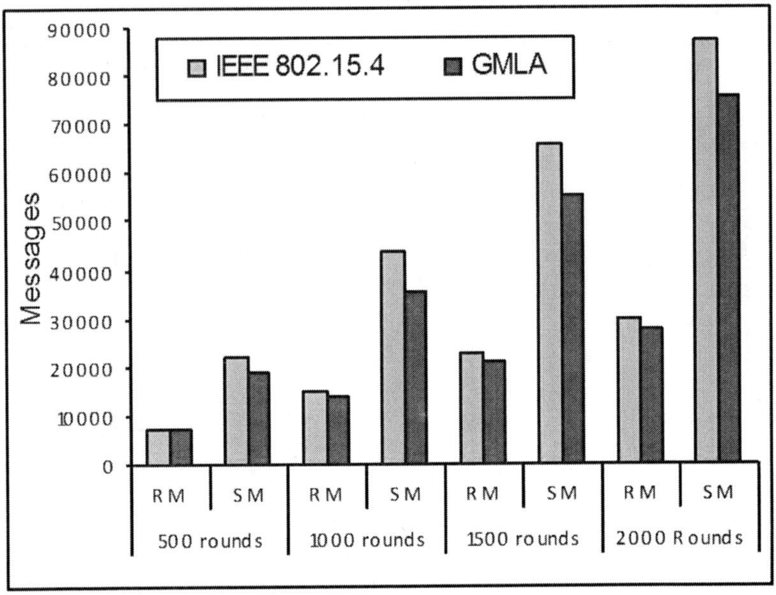


**Figure 7.** Sent (SM) and Received Messages (RM).

The VOA algorithm was assessed with the help of an experimental setup based on real situations and one of the experiments has been performed by varying the number of slave nodes. The results are shown in Figure 8. The goal was to evaluate the influence of the number nodes on the *Ef* and *QoF*metrics. When compared with VOA, IEEE 802.15.4 presents similar results for just one case: a network with 4 slaves. When the number of slaves increases, the difference between VOA and IEEE 802.15.4 become greater. The difference of efficiency between VOA and IEEE 802.15.4 when considering 29 slaves is of more than 100%. These results show that VOA has a satisfactory performance and maintains a minimum QoS level even with a high number of slaves (Pinto e Montez, 2010).

The Decentralized Power-Aware Wireless Sensor Network (DPAWSN) approach has the main goal of maintaining a minimum *QoF*, while improving the *Ef* and saving energy. This approach can be considered as decentralized, due to the fact that nodes have autonomy to decide whether to send or not to send messages. On the one hand, a certain *QoF* level is imposed by network administrator, on the other hand, the WSN´s lifetime is increased by the power awareness decision taken in each node.

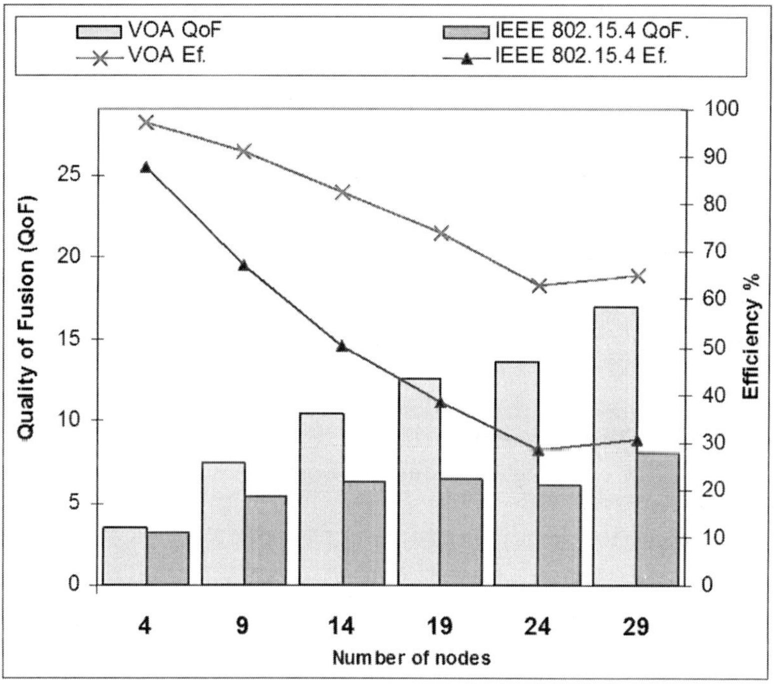


**Figure 8.** Comparison of VOA and IEEE 802.15.4 protocol with variable number of nodes.

The main idea behind DPAWSN is that the base station will control each node in order to decreases or increases the transmission rate when the *QoF* level is above or below the target value. Thus, DPAWSN is able to maintain a *QoF* level and increase the WSN´s lifetime due to the fact that nodes present a selfish behaviour as the nodes transmission rate is calculated based on individual remaining voltage.

The test assessment was conducted in a noisy environment with an unspecified number of computers communicating using IEEE 802.11.

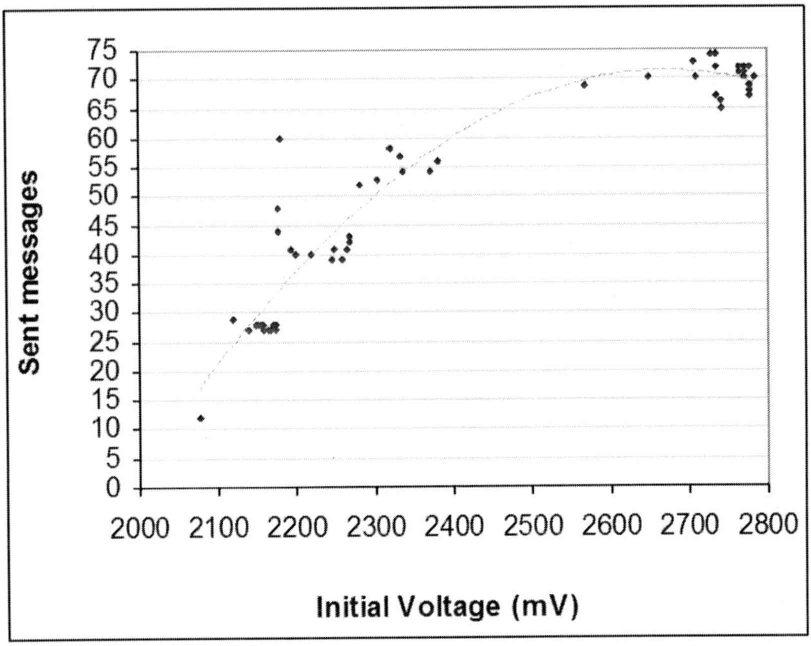


**Figure 9.** Dispersion Graphic of 30 tests (each point represents sent messages of lower or higher battery level nodes).

Figure 9 shows the correlation between sent messages and the initial voltage in each node. Note that the initial value on the x-axis is 2000mV, which represents the minimum battery conditions for a node to work. It is possible to notice that two regions are present in the graph: the first one represents sent messages of nodes with a lower level of battery and the second one represents sent messages of nodes with a higher level of battery. In more detail, each point represents the average number of sent messages (y-axis) over the initial voltage average of the node (x-axis). The effectiveness of DPAWSN is confirmed by Figure 9, since lower battery level are applied to nodes that send less messages and nodes with higher battery level do send more messages.

Finally, DPAWSN is able to autonomically adjust the transmission rate based on the voltage level of the nodes. Moreover, it is able to achieve a targeted *QoF* imposed by system administrators.

# 7. FINAL CONSIDERATIONS AND FUTURE DIRECTIONS

WSNs represent one of the most interesting solutions to monitor and sense data in hazardous or inaccessible areas. This Chapter presented various possible approaches that designers might adopt in order to provide an energy efficient WSN. Due to the enormous variety of environments that these networks may be used in, numerous challenges and constraints have to be considered when choosing the optimization approach. Also, different goals and applications call for different targets that may not be achieved up to a satisfying level at the same time. Therefore, some approaches offer to define a trade-off between opposing aims, such as *QoS* and energy consumption. Each solution presented in this Chapter is aiming at solving at least one identified cause of extensive energy consumption. The approaches are using the WSN's different design levels to increase the network's lifetime, *QoS*, and/or optimize other points that may or may not suit the designer.

Taking into account that lifetime maximization is one of the main goals of WSN approaches, and as the WSN devices consume more energy during the transmission and reception of packets, even in short distances, than other tasks such as those related to processing, sensing and data storing, the design of efficient protocols and communication algorithms is a research direction.

The current tendency goes towards solutions that involve a trade-off between more than one constraint or that may adapt or change the behaviour of the WSN's nodes during its employment are showing that researches are aware of the complexity and unpredictability of the environment and task of such networks.

However, some existing research areas are becoming more relevant, mainly due to the recent technology evolution of these networks. For example, the most popular motes in the years 2000-1010 were based on 4 to 8 MHz processors and 128 Kbytes memory; but recently there are motes with 180MHz processor and up to 4Mbytes memory, supporting Java Virtual Machine. Therefore, the hardware evolution trend directs researches to implement more sophisticated and robust approaches in an autonomic and distributed way with multi-objective optimization approaches, however with power consumption as an important goal.

The gradual replacement of very expensive centralized sensor systems by a set of wireless sensor nodes, which operate in a collaborative and autonomic way (mainly with self-management and self-healing properties) is also becoming a trend. Thus, multi-agent approaches and lightweight optimization techniques are emerging as an alternative. This fact is mainly due to the distributed and optimized way that these approaches perform.

Moreover, the gradual increase in the motes´ local storage capacity has induced the use of WSN data mules. The respective research focuses on the development of architectures and algorithms where nodes must locally storage their sensed data until mobile nodes (mobile base stations) gather this information.

As final consideration it is to be stated that the variety of challenges has generated an even greater number of approaches to deal with the concerns that WSNs face in all the possible harsh and noisy environments they may face today. It is now the designers' task to find the best match or combination to optimize the network according to its environment, its tasks, and its most important requirements and constraints. As there is no solution that may cover all problems at the same time, the correct analysis of problems to be expected has become one of the most important parts of the work of today's designers.

## NOTES

[1] - freely available at http://www.control.lht.se/truetime.

## REFERENCES

1. I. F. Akyildiz, W. Su, Y. Sankarasubramaniam, E. Cayirci, 2002A Survey on Sensor Networks. IEEE Communications Magazine, 102114

2. I. F. Akyildiz, F. Brunetti, C. Blazquez, 2008Nanonetworks: A new communication paradigm. Computer Networks, 5222602279

3. C. Allipi, C. Galperti, 2008An Adaptive System for Optimal Solar Energy Harvesting in Wireless Sensor Networks Nodes. IEEE Transactions on Circuits and Systems, 55617421750

4. G. Campobello, A. Leonardi, S. Palazzo, 2012Improving Energy Saving and Reliability in Wireless Sensor Networks Using a Simple CRT-Based Packet-Forwarding Solution. IEEE/ACM Transactions on Networking, 201

5. W. Chen, M. J. Neely, U. Mitra, 2008Energy-Efficient Transmissions With Individual Packet Delay Constraints. IEEE Transactions on Information Theory, 545

6. P. H. Chou, C. Park, 2005Energy Efficient Platform Designs for Real-World Wireless Sensing Application. Proceedings of IEE/ACM International Conference Computer-Aided Design

7. A. Costa, A. Sayeed, 2003Data Versus Decision Fusion for Distributed Classification in Sensor Networks. Proceedings of the IEEE International Conference on Acoustics, Speech, and Signal Processing, 832835

8. I. Demirkol, C. Ersoy, F. Alagöz, 2006MAC Protocols for Wireless Sensor Networks: A Survey, IEEE Communications Magazine

9. L. Deying, J. Xiaohua, L. Hai, 2004Energy Efficient Broadcast Routing in Static Ad Hoc Wireless Networks. IEEE Transactions on Mobile Computing, 32

10. E. Felemban, C. Lee, E. Ekicin, 2006MMSPEED: Multipath Multi-SPEED Protocol for QoS Guarantee of Reliability and Timeliness in Wireless Sensor Networks. IEEE Transactions on Mobile Computing, 56738754

11. B. Gholamzadeh, H. Nabovati, 2008Concepts for Designing Low Power Wireless Sensor Networks, World Academy of Science, Engineering and Technology

12. S. Guha, P. Basu, C. Chau, R. Gibbens, 2011Green Wave Sleep Scheduling: Optimizing Latency and Throughput in Duty Cycling Wireless Networks. IEEE Journal on Selected Areas in Communications, 298

13. S. Hadim, N. Mohamed, 2006Middleware Challenges and Approaches for Wireless Sensor Networks. IEEE Distributed Systems Online, 73

14. I. Ho, S. Liew, 2007Impact of power control on performance of IEEE 802.11 wireless networks. IEEE Trans. Mobile Comput., 61112 451258November 2007

15. J. Hong, I. Jang, H. Lee, S. Yang, H. Yoon, 2010MRMAC: Medium Reservation MAC Protocol for Reducing End-to-End Delay and Energy Consumption in Wireless Sensor Networks. IEEE Communications Letters, 147

16. M. C. Huebscher, J. A. Mc Cann, 2008A survey of autonomic computing degrees, models, and applications. ACM Comput Surveys

17. IEEE.2008Especificação IEEE 802.15.4, Available from : http://www.ieee802.org/15/pub/TG4.html

18. M. Ilyas, I. Mahgoub, 2005Handbook of Sensor Networks: Compact Wireless and Wired Sensing Systems, CRC Press Inc., Boca Raton, USA

19. S. Jungmin, K. Jintae, G. Indanil, 2005Cushion: Autonomically Adaptative Data Fusion in Wireless Sensor Networks, Proceedings of MASS'05

20. H. Karl, A. Willig, 2005Protocols and Architectures for Wireless Sensor Networks, John Willey & Sons

21. Kephart, J.O., Chess, D.M., The Vision of Autonomic Computing, IEEE Computer, 2003.

22. S. Kim, S. Pakzad, D. Culler, J. Demmel, G. Fenves, S. Glaser, M. Turon, 2007Health Monitoring of Civil Infrastructures Using Wireless Sensor Networks. Proceedings of the 6th International Conference on Information Processing in Sensor Networks, Massachusetts, USA

23. J. Kim, Y. Know, 2008Interference-Aware Transmission Power Control for Wireless Sensor Networks. IEICE Trans. Commun., E91-B11November 2008

24. M. Kim, Y. Han, H. Park, 2011Energy-Aware Hybrid Data Aggregation Mechanism Considering the Energy Hole Problem in Asynchronous MAC-Based WSNs. IEEE Communications Letters, 1511November 2011

25. A. Koubaa, R. Severino, M. Alves, E. Tovar, 2009Improving Quality-of-Service in Wireless Sensor Networks by mitigating "Hidden-Node Collisions". IEEE Transactions on Industrial Informatics, Special Issue on Real-Time and Embedded Networked Systems, 53August 2009

26. M. Kubisch, H. Karl, A. Wolisz, L. C. Zhong, J. Rabaey, 2003Distributed Algorithms for Transmission Power Control in Wireless Sensor Networks, Proceedings of WCNS03, IEEE Wireless Communications and Networking Conference, March 2003

27. F. Lavratti, A. Ceratti, D. Prestes, L. Bolzani, F. Vargas, C. Montez, F. Hernandez, E. Gatti, 2012A Transmission Power Self-Optimization Technique for Wireless Sensor Networks. ISRN Communications and Networking.

28. W. L. Lee, A. Datta, R. Cardell-Oliver, 2008FlexiTP: A Flexible-Schedule-Based TDMA Protocol for Fault-Tolerant and Energy-Efficient Wireless Sensor Networks. IEEE Transactions on Parallel and Distributed Systems, 196

29. S. Lin, J. Zhang, G. Zhou, L. Gu, T. He, J. A. Stankovic, 2006ATPC: Adaptive Transmission Power Control for Wireless Sensor Networks. Proceedings of SenSys'06, November 2006

30. T. Melodia, D. Pompili, I. F. Akyildiz, 2010Handling Mobility in Wireless Sensor and Actor Networks. IEEE Transactions on Mobile Computing, 92160173

31. Z. Mica, Datasheets. Mote, 2012Available from: www.xbow.com

32. M. Molla, S. I. Ahamed, A. Survey, Sensor. for, Network, Challenges, 2006Proceedings of the 2006 International Conference on Parallel Processing Workshops, 2006

33. J. P. Monks, W. Bharghawan, W. Mei, W. Hwu, 2001Power Controlled Multiple Access Protocol for Wireless Packet Networks. Proceedings of INFOCOM, 219228

34. K. Morita, K. Watanabe, N. Hayashibara, T. Enokido, M. Takanizawa, 2007Efficient Data Transmission in a Loss and Resource Limited Wireless Sensor-Actuator Network. Proceedings of the 10th IEEE International Symposium on Object and Component-Oriented Real-Time Distributed Computing, 2007

35. R. S. Oliver, G. Fohler, 2010Timeliness in Wireless Sensor Networks: Common Misconceptions. Proceedings of the 9th International Workshop on Real-Time Networks, Brussels, Belgium, July 2010

36. O. Omeni, A. C. W. Wong, A. J. Burdett, C. Toumazou, 2008Energy Efficient Medium Access Protocol for Wireless Medical Body Area Sensor Networks. IEEE Transactions on Biomedical Circuits and Systems, 24December 2008

37. B. Otal, L. Alonso, C. Verikoukis, 2009Highly Reliable Energy-Saving MAC for Wireless Body Sensor Networks in Healthcare Systems. IEEE Journal on Selected Areas in Communications, 274

38. S. Patil, S. K. Das, A. Nasipuri, 2004Serial Data Fusion Using Space-filling Curves in Wireless Sensor Networks. Proceeding of IEEE SECON, 182190

39. A. R. Pinto, C. Montez, Autonomic Approaches for Enhancing Communication QoS in Dense Wireless Sensor Networks with Real Time Requirements, 2010IEEE Int. Test Conference, 2010. 110

40. Z. Quan, A. Subramanian, A. H. Sayed, 2007REACA: An Efficient Protocol Architecture for Large Scale Sensor Networks. IEEE Transactions on Wireless Communications, 61038463855

41. W. Raghunathan, S. Gareniwal, B. Srivastava, 2006Emerging Techniques for Long Lived Wireless Sensor Networks. IEEE Communications Magazine, 444108114April 2006.

42. M. Rossi, M. Zorzi, 2007Integrated Cost-Based MAC and Routing Techniques for Hop Count Forwarding in Wireless Sensor Networks, IEEE Transactions on Mobile Computing, 64April 2007

43. L. Shi, A. O. Fapojuwo, 2010TDMA Scheduling with Optimized Energy Efficiency and Minimum Delay in Clustered Wireless Sensor Networks. IEEE Transactions on Mobile Computing, 97July 2010

44. K. Sohraby, D. Minoli, T. Znati, 2007Wireless Sensor Networks: Technology, Protocols and Applications, John Wiley & Sons

45. J. A. Stankovic, T. F. Abdelzaher, C. Lu, L. Sha, J. C. Hou, 2003Real-Time Communication and Coordination in Embedded Sensor Networks. Proceedings of the IEEE, 91710021022

46. H. Tseng, S. Sheu, Y. Shih, 2011Rotational Listening Strategy for IEEE 802.15.4 Wireless Body Networks. IEEE Sensors Journal, 119

47. Q. Wu, N. S. V. Rao, J. Barhen, S. S. Iyenger, V. K. Vaishnavi, H. Qi, K. Chakrabarty, 2004On Computing Mobile Agent Routes for Data Fusion in Distributed Sensor Networks. IEEE Transactions on Knowledge and Data Engineering, 166

48. X. Wu, G. Chen, S. K. Das, 2008Avoiding Energy Holes in Wireless Sensor Networks with Nonuniform Node Distribution. IEEE Transactions on Parallel and Distributed Systems, 195

49. Y. Wu, S. Fahmy, N. Shroff, 2009Optimal Sleep/Wake Scheduling for Time-Synchronized Sensor Networks With QoS Guarantees. IEEE/ACM Transactions on Networking, 17515081521

50. Y. Wu, X. Li, Y. Liu, W. Lou, 2010Energy-Efficient Wake-Up Scheduling for Data Collection and Aggregation. IEEE Transactions on Parallel and Distributed Systems, 212

51. J. Yick, B. Mukherjee, D. Ghosal, 2008Wireless sensor network survey. Computer Networks, 5222922330

52. L. Zhao, L. Guo, J. Zhang, H. Zhang, 2009Game-theoretic medium access control protocol for wireless sensor networks. IET Communication, 3812741283

3

# SUB-SAMPLING FRAMEWORK COMPARISON FOR LOW-POWER DATA GATHERING: A COMPARATIVE ANALYSIS

**Bojan Milosevic [1,2,*], Carlo Caione [1], Elisabetta Farella [1,2], Davide Brunelli [3] and Luca Benini [1,4]**

[1]DEI, University of Bologna, 40123 Bologna, Italy

[2]E3DA, Fondazione Bruno Kessler, I-38123 Trento, Italy

[3]DII, University of Trento, I-38123 Trento, Italy

[4]IIS, ETH, 8092 Zurich, Switzerland

## ABSTRACT

A key design challenge for successful wireless sensor network (WSN) deployment is a good balance between the collected data resolution and the overall energy consumption. In this paper, we present a WSN solution developed to efficiently satisfy the requirements for long-term monitoring of a historical building. The hardware of the sensor nodes and the network deployment are described and used to collect the data. To improve the network's energy efficiency, we developed and compared two approaches, sharing similar sub-sampling strategies and data reconstruction assumptions: one is based on compressive sensing (CS) and the second is a custom data-driven latent variable-based statistical model (LV). Both approaches take advantage of the multivariate nature of the data collected by a heterogeneous sensor network and reduce the sampling frequency at sub-Nyquist levels. Our comparative analysis highlights the advantages and limitations: signal reconstruction performance is assessed jointly with network-level energy reduction. The performed experiments include detailed performance and energy measurements on the deployed network and explore how the different parameters can affect the overall data accuracy and the energy consumption. The results show how the CS approach achieves better reconstruction accuracy and overall efficiency, with the exception of cases with really aggressive sub-sampling policies.

## KEYWORDS

wireless sensor networks; low-power design; data reconstruction; compressive sensing; latent variables

## 1. INTRODUCTION

The recent technological evolution of sensing devices for wireless sensor networks (WSNs) has triggered research activities in the field of data gathering and compression. The goal is to minimize sampling, storage and communication costs to extend as much as possible the lifetime of the network, while maintaining the desired data accuracy [1].

In a typical WSN scenario, the communication pattern consists of a very large number of small and low-cost devices that periodically sample the sensors' data and transmit them to a central collecting sink [2]. For large scenarios, where the device transmission range is smaller than the area to cover, multi-hop communication is used to efficiently gather the desired data. In this case, the network topology organization and the data transmission and routing approaches should be optimized to ensure the correct network coverage and connectivity and to balance data acquisition quality and power consumption [3].

Common approaches to extend the uptime of a sensor node try to enhance the battery life directly by harvesting energy from the environment and employing low-power hardware architectures [4] or using improved wireless protocols and distributed data processing [5]. More recently, researchers have been optimizing the battery life indirectly by reducing the overall amount of data sensed and transmitted through the network [6,7]. When the considered signals present spatio-temporal correlations, those can be used to reconstruct the desired data from only a limited portion of collected samples [8]. The ability to reconstruct missing data enables the adoption of aggressive duty cycling policies on individual sensor nodes, sampling only a limited part of the data, thereby reducing the energy consumption [9].

Among the different proposed approaches to increase the energy efficiency of a sensor network, we focus on strategies that leverage correlations to reconstruct sub-sampled signals. Two of the most promising techniques capable of successfully recovering the signal from highly incomplete versions are compressive sensing (CS) [10,11] and a data-driven statistical model based on latent variables (LV) [12,13]. These techniques are both able to recover the original signals from a smaller sub-sampled version obtained by skipping samples during the acquisition phase.

CS theory claims that if a signal can be compressed using classical transform coding techniques and its representation is sparse on some basis, then a small number of projections on random vectors contains enough information for approximate reconstruction [10]. Natural signals have usually a relatively low information content as measured by the sparsity of their spectrum [14]; therefore, the theory of CS suggests that randomized low-rate sampling may provide an efficient alternative to high-rate uniform sampling. This peculiar form of CS is a novel strategy to sample and process sparse signals at a sub-Nyquist rate [15].

The second considered framework is an energy-efficient and data-driven technique to estimate missing data within a heterogeneous sensor network, based on latent variables [12]. This approach extends the standard latent variable factorization model [16], which typically considers only dyadic interactions in data, to multivariate spatio-temporal data, by applying tensor decomposition techniques [17]. The key advantage of using a latent variable model is that it provides a compact representation of the gathered data that can be used to recover the missing samples. To perform well under extreme sub-sampling conditions, the standard technique is extended to explicitly incorporate the spatial, temporal and inter-sensor correlations present in the considered data [13].

In this work, we focus on the comparison between these two approaches and their impact on the energy consumption of the network nodes. We describe a real deployment of a WSN for the environmental monitoring of a heritage building, which is used as a common platform and testbed for the evaluation. The two techniques share the same approach, but address the reconstruction problem from different theoretical bases; hence, a straightforward question is which one is most well suited as an energy-minimization technique for WSNs. To the best of our knowledge, this is the first work in which the two techniques are directly compared, using the same dataset and a detailed power consumption model to evaluate their reconstruction abilities and the energy efficiency.

To compare the performance of the two techniques, evaluating the best trade-off between signal recovery and energy savings, we explore the impact of features, such as the use of network-wide correlations to improve the reconstruction accuracy or the length of the data block to be reconstructed. For both the hardware characteristics and the gathered data, we use a real deployed sensor network. The power consumption measurements are used to model the node's power consumptions to allow an easy evaluation of the energy expenditure for the different algorithms and sampling strategies used.

The rest of the paper is organized as follows: In Section 2, the related works are introduced. In Sections 3 and 4, the mathematical background for CS and LV is presented, while in Section 5 the models for the hardware, network and the power consumption are described. The simulation results are shown in Section 6, and the conclusions are in Section 7.

## 2. RELATED WORK

The problem of data gathering, compression and signal reconstruction in WSNs is well explored in the literature. Even though the majority of the works deal with reconstruction algorithms and mathematical aspects, practical aspects and low power implementation problems have lately been gaining interest.

### 2.1. Theoretical Approaches

As seen in the previous section, CS builds on several works, like [10,11], which show that if a signal can be compressed using classical transform coding techniques and its representation is sparse on some basis, then a small number of projections on random vectors contain enough information for approximate reconstruction.

For WSN data gathering applications, CS can be used to improve the overall efficiency of the network. One approach is to use CS to aggregate the data as it is collected by a multi-hop network, avoiding a progressive increase of data to be transmitted as we travel from the peripheral nodes to the sink. The work in [18] introduces an approach to minimize the network energy consumption through joint routing and compressed aggregation. The approach has an optimal solution, which is NP-complete, so the authors present a greedy heuristic that delivers near-optimal solutions for large-scale WSNs. Another data acquisition protocol for WSNs is presented in [19], where the results from studies in vehicle routing are applied to data collection in WSN and integrated with CS. Furthermore, this approach is NP-hard, and the authors propose both a centralized and a distributed heuristic to reduce computational times and to improve the scalability of the approach. These works analyze the routing and aggregation problem in WSNs, while in our work, we focus on the spatio-temporal correlation in the observed phenomena, which can be used to improve the data reconstruction accuracy Our approach allows the sensor nodes to duty-cycle and skip a significant amount of samples, thus saving energy Our analysis focuses on joint data reconstruction and on power consumption impacts in a real deployment scenario.

When CS is used in a scenario where several sensors acquire data from the same environment, we can think that the sensed data have a certain kind of shared information that can be exploited to perform a better reconstruction. The best known technique used for exploiting the inter- and intra-correlation among several nodes in a WSN is the distributed CS (DCS) introduced in [20]. In these works, the authors analyze three different Joint Sparsity Models (JSMs): JSM-1, JSM-2 and JSM-3, to describe most of the signal ensemble; and for each sparsity model, they present a different reconstruction algorithm. Among other works dealing with joint sparse recovery, we can cite [21] or Kronecker CS, introduced in [22]. Different from these works, in this paper, we want to focus on a special technique for the reconstruction of jointly sparse solutions, known as multiple measurement vector (MMV), based on $\ell_{2,1}$-regularization, which is introduced in [23].

Latent variables and decomposition-based techniques have also been proposed in the literature, with their most notable applications in collaborative filtering [24] and recommender systems [25]. A standard way to learn a set of latent variables for a two-dimensional dataset is to apply a matrix factorization technique. A recent review of matrix factorization techniques with applications to recommender systems can be found in [16]. To achieve a better latent variable model and to allow the prediction of the missing entries in the original data matrix, standard matrix factorization techniques have been extended to incorporate temporal dynamics and, thus, better capture the temporal evolution of the data [12,26]. When dealing with multivariate datasets, which can be represented by $N$-dimensional data arrays, tensor factorization techniques can be exploited to further improve the performance of the models [27]. In our work, we applied a tensor factorization approach to WSNs, taking advantage of the three-dimensional nature of the data gathered from a heterogeneous sensor network. We further extended this approach, incorporating for each data dimension a correlation model learned from the data itself, to enhance the capabilities to reconstruct the missing data.

## 2.2. WSN-Related Practical Implementations

The general problem of using CS in WSNs is investigated in several works. In [28], the measurement matrix is created jointly with data routing policies, trying to achieve an efficient reconstruction quality. Furthermore, in [29], the authors try to improve the reconstruction by reordering the input data to achieve a better compressibility. The main focus of these works is to investigate the signal reconstruction problem, but they are missing the consideration about how the usage of CS impacts the power consumption. While there is no doubt that CS is a powerful technique for data size reduction and compression, its usage and impact on network lifetime, when real hardware and COTSnodes are used, is still marginally addressed in literature.

The work in [30] is one of the first papers trying to address the issue of energy consumption for data compression, dealing with the problem of generating a good measurement matrix using as low energy as possible. In this work, the research is focused on wireless body sensor networks (WBSNs) for real-time energy-efficient ECG compression. This is quite a different application when compared to WSNs, where the presence of several nodes, sensing the same environment, permits one to exploit the distributed nature of the signals to improve the quality of the data recovery.

Other works deal with the use of CS when data are gathered from different joint sources, using DCS to improve the recovery quality. In [31], DCS and principal component analysis are used to reconstruct spatially- and temporally-correlated signals in a sensor network, but once again, the contribution of the power consumption for compression in the network lifetime is neglected.

In this paper, we deal with CS when the signals are sampled at a sub-Nyquist frequency, resembling a technique that is referred to analog CS in the literature. The name derives from the fact that the sub-sampling is performed by dropping samples during the acquisition and analog-to-digital conversion (ADC) stage [32]. One notable example of this technique is in [33], where the effects of circuit imperfections in the analog compressive sensing architectures are discussed. In the framework proposed in this work, samples are not discarded by analogue circuits, but are not sampled at all, saving the energy for waking up the node.

In the literature, other works investigate the problem when the samples are discarded by the device performing the sensing rather than the ADC. For example, in [34,35], the analysis of energy consumption is totally neglected, and the recovery is strictly related to the specific applications described. Different from environmental signals, used in our work, the signals coming from the oximeter present a much higher temporal correlation, with small variations in their temporal evolution, facilitating their reconstruction.

Matrix factorization learning of latent variables has been used for recovering missing data in sensor networks in [12], where temporal correlations found in the dataset are used to infer the missing variables. Tensor decomposition techniques have been applied on WSNs in [36], where the learned models are used to find the damage in a structural health monitoring application. The previously presented algorithms only consider homogeneous sensor streams, dealing with one sensor at a time, and do not consider the energy costs across the network. Instead, our approach focuses on the multivariate nature of the collected data, and it expands the tensor factorization techniques by employing spatio-temporal and intra-sensor correlations for more robust and better results than the existing methods.

The energy consumption during sub-sampling operations is instead considered in [37], where the authors use a sparse generated matrix adjusting the sampling rate to maintain an acceptable reconstruction performance, while minimizing the energy consumption. Different from our work, they do not consider inter- and intra-correlation among signals and do not use any group sparsity-enhancing algorithm to perform a better recovery. In addition, our work presents a real network deployment and takes advantage of the gathered data and power consumption measurements to accurately evaluate the two proposed frameworks in a common testbed.

# 3. COMPRESSING SENSING AND GROUP SPARSITY

## 3.1. MATHEMATICAL BACKGROUND

Considered a continuous signal $x(t)$ of duration $T$, and $x(n)$ $1 \leq n \leq N$ is its discrete version. The Nyquist theorem states that in order to perfectly capture the information of the signal $x(t)$, having a bandwidth of $B_{\mathrm{nyq}}/2$ Hz, we must sample the signal at its Nyquist rate of $B_{\mathrm{nyq}}$ samples per second.

In formula:

$$x(n) = x(t)|_{t=nT_s} \tag{1}$$

where $T_s \leq 1/B_{\mathrm{nyq}}$ and $NT_S < T$. The sampled signal $x(n)$ is represented by an $N$-dimensional vector of real numbers $\mathbf{x}$.

If the vector $\mathbf{x}$ is sparse, then CS is able to recover this finite-dimensional vector $\mathbf{x} \in \mathbb{R}^N$ from a very limited number of measurements of the original signal $x(t)$. The sparsity of a signal $\boldsymbol{\alpha}$ is usually indicated as the $\ell_0$-norm of the signal, where the $\ell_p$-norm $\|\cdot\|_p$ is defined as:

$$\|\alpha\|_p = \left( \sum_{i=0}^{N-1} |\alpha_i|^p \right)^{1/p} \tag{2}$$

with $\boldsymbol{\alpha} \in \mathbb{R}^N$. If the signal $\mathbf{x}$ is sparse, then there exists some $N \times N$ basis or dictionary $\boldsymbol{\Psi} \in \mathbb{R}^{N \times N}$, such that there is an $N$-dimensional vector $\boldsymbol{\alpha}$ implying $\mathbf{x} = \boldsymbol{\Psi}\boldsymbol{\alpha}$ and $\|\boldsymbol{\alpha}\|_0 \leq K$ with $K \ll N$. CS theory demonstrates that it is possible to compress this kind of sparse signal using a second different measurement matrix $\boldsymbol{\Phi} \in \mathbb{R}^{M \times N}$ with $M \ll N$ [11]. The compression can be written as $\mathbf{y} = \boldsymbol{\Phi}\mathbf{x}$, where $\mathbf{y}$ is the $M$-dimensional measurements vector. While **Ψ** is usually defined by the signals characteristics, **Φ** has to be designed so that *M* is much smaller than *N*.

Having the measurements vector **y**, the recovery of the original signal **x** can be obtained by the inverse of the measurement problem:

$$y = \Theta\alpha = \Phi\Psi\alpha \tag{3}$$

Even though the inversion is not an easy task, since the matrix $\mathbf{\Theta} \in \mathbb{R}^{M\times N}$ is rectangular with $M \ll N$, the fact that **x** is sparse can relax the problem by opening the way to the use of optimization-based reconstruction or iterative support-guessing reconstruction.

The most common optimization-based method, here reported for the sake of clarity, is the basis pursuit (BP) [38] method that searches for the most sparse solution for which the $\|\boldsymbol{\alpha}\|_1$ is minimum:

$$\hat{\alpha} = \arg\min \|\alpha\|_1 \text{ subject to } y = \Theta\alpha = \Phi\Psi\alpha \tag{4}$$

CS theory proves that if the two matrices **Φ** and **Ψ** are incoherent (elements of the matrix **Φ** are not sparsely represented in the basis **Ψ**) and the original signal x is compressible or sparse, we can recover $\boldsymbol{\alpha}$ with high probability.

To further enhance the recoverability, recent studies propose taking into account additional information about the underlying structure of the solutions [39]. When the signals to compress and recover are obtained from sensors deployed close to each other in the environment, we can expect that the ensemble of these signals presents an underlying joint structure. This characteristic can be exploited to further compress the data, without a loss in reconstruction accuracy In practice, this class of solutions is known to have a certain group sparsity structure. This means that the solution has a natural grouping of its components, and the components within a group are likely to be either all zeros or all non-zeros. Encoding the group sparsity structure can reduce the degrees of freedom in the solution, thereby leading to better recovery performance.

Having an ensemble of *J* signals, we can denote each signal with $\mathbf{x}_j \in \mathbb{R}^N$ with $j \in \{1,2,..., J\}$. For each signal $\mathbf{x}_j$ in the ensemble, we have a sparsifying basis $\mathbf{\Psi} \in \mathbb{R}^{N\times N}$ and a measurement matrix $\mathbf{\Phi}_j \in \mathbb{R}^{M\times N}$, such that, as before, $\mathbf{y}_j = \mathbf{\Phi}_j\mathbf{x}_j$ with$M_j \ll N$ and $\mathbf{x}_j = \mathbf{\Psi}\boldsymbol{\alpha}_j$. The reconstruction of jointly sparse solutions, also known as the multiple measurement vector

(MMV) problem, has its origin in sensor array signal processing and recently has received much interest as an extension of the single sparse solution recovery in compressive sensing. The recovery problem can be formulated as:

$$\min_{\tilde{\alpha}} \quad \|\tilde{\alpha}\|_{w,2,1} := \sum_{i=1}^{n} w_i \|\tilde{\alpha}_i\|_2 \qquad (5)$$
$$\text{subject to} \quad \tilde{\Theta}\tilde{\alpha} = \tilde{Y}$$

where $\tilde{Y} = [y_1^T y_2^T \ldots y_J^T]^T$, $\tilde{\alpha} = [\alpha_1^T \alpha_2^T \ldots \alpha_J^T]^T$, $w_i$ is the weight and $\tilde{\boldsymbol{\Theta}} \in \mathbb{R}^{JM \times JN}$ is a matrix having on the diagonal matrices $\Theta_j = \Phi_j \Psi$ for $j \in \{1, 2, \ldots, J\}$.

## 3.2. CS and Sub-Nyquist Sampling

As discussed in the previous section, to successfully recover the original signal from its sampled version, the samples are taken regularly on a time axis at a given rate that is not less than the Nyquist one. With respect to CS, this requirement means that the measurement matrix $\boldsymbol{\Phi}$ is a dense matrix (usually an independent and identically distributed Gaussian matrix). A particular form of CS, usually referred to as analog CS, relies on random sampling and aims to produce a number of measurements fewer than with the Nyquist sampling, still enabling the reconstruction of the original signal. While analog CS is usually performed by means of specialized hardware encoders, this is also a suitable technique to be performed on WSNs nodes, opportunely skipping samples during the acquisition phase.

From a mathematical point of view, the problem is still the same as the problem in Equation (3); what is different is the form of the measurement matrix $\boldsymbol{\Phi}$, which is not a dense matrix, but is a sparse measurement matrix. More precisely, if $\mathbf{B}$ is an $M$-dimensional vector where each element is a unique random entry between one and $N$, then the matrix $\boldsymbol{\Phi}$ in the analog CS is a sparse $M \times N$ measurement matrix, which is composed by an all-zero vector on each row and a "1" at the location given by the $i$-th element of $\mathbf{B}$. This is a very simple measurement matrix, energetically inexpensive to generate and store and permits a huge reduction in the duty-cycling of the nodes.

# 4. LATENT VARIABLES AND TENSOR FACTORIZATION

Latent variable-based factorization is a simple, yet powerful framework for modeling data and has been successfully applied in several application domains [16]. The main idea behind this framework is to model the large number of observed variables (the observed data) in terms of a much smaller number of unobserved variables (the latent variables). The latent variables are learned from the observed data and are used to estimate the missing samples, modeling complex interactions between the observed variables through simple interactions between the latent variables.

More specifically, given some multivariate data that are collected by a heterogeneous WSN in a large field over time, we can naturally organize these into a three-dimensional data array (or a three-tensor). Each of the three dimensions corresponds to a different variate of a particular measurement (e.g., the time, the location and the sensor type associated with each reading). Once the data are organized in this way, we can then associate a low-dimensional latent variable with each unique location, time slice and sensor type. We can thus model a particular observation (at a given location, time and type) as a noisy combination of the associated latent variables. In many scenarios, a multiplicative combination of these latent variables is able to capture intricate dependencies in the data [17,40]. The goal then is to learn a good set of latent variables (that is, find a factorization) that can efficiently represent our observed data.

## 4.1. Modeling Details

We now formally define our model. For each unique time instance $t$, sensor type $s$ and node location $n$, we associate a unique $K$-dimensional vector, $a_t$, $b_s$ and $c_n$, respectively. These unobserved vectors are called the latent factors or variables and are assumed to control the location-, time- and sensor-specific interactions present in the observed data.

Then, given a $[T \times S \times N]$ tensor $\chi$ of sensor readings from $S$ different sensor types collected at $N$ different nodes and $T$different time instances, with possible missing entries, we model $\chi$ as follows (see Figure 1). We assume that each reading$x_{tsn}$ (the reading at time $t$, for sensor type $s$, at node location $n$) is a noisy realization of the underlying true reading that is obtained by the interaction of the time-specific latent variable $a_t$, with the sensor-specific latent variable $b_s$ and with the location-specific variable $c_n$.

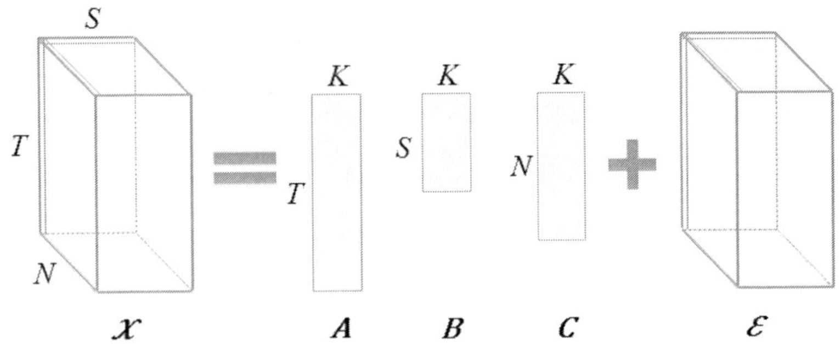


**Figure 1.** Tensor data representation and its matrix factorization.

$$x_{tsn} = \sum_{k=1}^{K} a_{tk} b_{sk} c_{nk} + \varepsilon \qquad (6)$$

where $\varepsilon$ is modeled as independent zero-mean Gaussian noise ($\varepsilon \sim \mathcal{N}(0, \sigma^2)$). Observe that under this model, once all of the latent variables are known, one can recover the true readings of all sensors at all locations and times. Thus, the goal is to find the most predictive set of vectors $a_t$, $b_s$ and $c_n$ for all $t = 1, \ldots, T$, $s = 1, \ldots, S$, $n = 1, \ldots, N$. Such a representation models the entire data of size $[T \cdot N \cdot S]$ by just $[K \cdot (T + N + S)]$ modeling parameters. The choice of the free parameter $K$ provides a key trade-off: a large $K$ increases the number of modeling parameters and, thus, can help model the observed data exactly. However, this lacks the capability to predict unobserved/missing data due to overfitting. A small $K$, on the other hand, escapes the overfitting problem, but the corresponding model lacks sufficient richness to capture salient data trends. The exact choice of a good $K$ is typically application dependent and is derived empirically.

### 4.1.1. Learning the Latent Variables

Finding the optimal set of $K$-dimensional latent variables given the observations is equivalent to factorizing the given tensor into three matrices, each of rank at most $K$ [17]. Thus, assuming that all of the data is known (that is, every entry in the tensor is observed), we can find the latent factors by employing the Canonical decomposition/Parallel factors (CP) tensor factorization [17]. This is simply a higher-order generalization of the matrix singular value decomposition (SVD) and decomposes a generic third order tensor in three matrix factors $A$, $B$ and $C$. Restricting the ranks of each of the matrix factors to at most $K$ yields the best rank $K$ approximation. Algorithmically, the matrix factors are found by an alternating least squares approach (ALS) [40], which starts from a random initialization and iteratively optimizes one matrix factor at a time, while keeping the other two fixed.

This technique can be generalized to work with tensors that have missing entries. Since sensor nodes can periodically go offline due to duty-cycling or running out of energy (preventing all sensors on a node from collecting any data for an extended period of time), we need to extend our basic model to deal with data missing from multiple sensors or nodes, resulting in entries and rows of missing data in the collected tensor. In order to do well in this regime, we extend the basic tensor factorization model to explicitly incorporate spatially-, temporally- and sensor-specific information from neighboring observations by explicitly learning and enforcing the corresponding correlations.

## 4.2. Incorporating Correlations

To successfully interpolate the sensor interactions to contiguously missing blocks of data, we need to explicitly model spatially-, temporally- and sensor-specific trends within each of our latent variables $a_t$, $b_s$ and $c_n$. Such trends ensure that the latent variables $a_t$ and $a_{t'}$ (respectively, $b_s$ and $b_{s'}$ and $c_n$ and $c_{n'}$) take similar values when times $t$ and $t'$ are "similar" (respectively, sensors types $s$ and $s'$ and locations $n$ and $n'$). Note that similarity can mean anything based on the context. For locations, it can mean that variables associated with two locations that are close in distance should have similar characteristics, while for time, it can mean that variables associated with times that are the same hour of the day or the same day of the week should have similar characteristics. Here, we take a data-driven approach to infer the best notion of similarity using correlations directly computed from the data.

We use a limited sample of data collected from the network to learn the correlations, which are then applied to the subsequent slot of data to reconstruct the unsampled values.

The similarity constraints are modeled in the same way for all three sets of latent variables, and here, we illustrate the case for the $a_t$'s. Since each $a_t$ is a $K$-dimensional variable, let $a_{tk}$ denote its $k$ – th coordinate. We model $a_{tk}$ (independently for each coordinate $k$) as:

$$\begin{aligned} a_{:k} &= \mu_{ak} + \alpha_{:k} \\ \alpha_{:k} &\sim \mathcal{N}(0, \Sigma_a) \end{aligned} \tag{7}$$

Here, $a_{:k}$ represents the collection of all $a_t$'s (across $t = 1,\dots,T$) in the $k$–th coordinate and $\mu_a$ represents their mean value. The distributional constraint over $\alpha_{:k}$ (as $\mathcal{N}(0,\Sigma_a)$) enforces the similarity constraints via the $T \times T$ covariance matrix $\Sigma_a$. By changing the $t$, $t'$ entry of $\Sigma_a$, we can encourage/discourage the corresponding $a_t$ and $a_{t'}$ to take similar values: a high positive value at $\Sigma_a(t, t')$ encourages a positive correlation; a high negative value encourages a negative correlation; while a value close to zero does not encourage any correlation. We enforce similar constraints on $b_s$ and $c_n$ variables, as well.

To get the right similarity constraints $\Sigma_a$, $\Sigma_b$ and $\Sigma_c$ (for latent variables $a_t$, $b_t$ and $c_n$), we compute them from the data that we are considering. For spatial similarity constrains, we computed the averaged pairwise Pearson correlation coefficients between data from different pairs of locations and averaged over different times and sensors. We do the same to approximate inter-sensor and temporal similarities. We can have only one global correlation constraint along each of the dimensions.

## 4.3. Parameter Learning

We can learn the underlying latent variables in a probabilistic framework using a maximum *a posteriori* (MAP) estimate. In particular, let $\theta$ denote all of the model parameters (*i.e.*, $\theta = \{\{a_t\}, \{b_s\}, \{c_n\}, \sigma\}$), then the optimum choice of parameters $\theta_{MAP}$given the data $\chi$ is obtained by:

$$\theta_{MAP}(\mathcal{X}) := \underset{\theta}{\operatorname{argmax}} \underbrace{p(\mathcal{X} \mid \theta)}_{\text{likelihood}} \underbrace{p(\theta)}_{\text{prior}} = \underset{\theta}{\operatorname{argmax}} \sum_{t,s,n \in \text{observed}} \log p(x_{tsn} \mid a_t, b_s, c_n, \sigma) +$$
$$+ \sum_{k=1}^{K} \log p(a_{:k}) + \sum_{k=1}^{K} \log p(b_{:k}) + \sum_{k=1}^{K} \log p(c_{:k}).$$

The first term (the likelihood) takes the form of Equation (6), and the other terms represent the priors for each latent variable; each one of them takes the form of Equation (7). We take a uniform prior over σ, the standard deviation of the residuals in Equation (6), so it does not explicitly show in the equation.

This optimization does not have a closed form solution. Standard gradient-based techniques can be used to get a locally optimal solution. Here, we can follow an alternating hill-climb approach by optimizing the value of one variable while keeping all others fixed to get a good solution. As for the standard factorization technique, this iterative approach uses a random initialization of the latent variables.

# 5. HARDWARE, NETWORK AND POWER MODELS

In this section, we will describe the hardware and software characteristics of the WSN deployed as a testbed for this work [41]. In particular, we report the experience gained during the EU Project 3ENCULT (Efficient energy for EU cultural heritage, http:// www. 3encult.eu) regarding the design and implementation of a WSN for environmental monitoring of heritage buildings. We present the network solution developed to efficiently satisfy the requirements for long-term monitoring of a historical building, called Palazzina della Viola, located in the center of Bologna, Italy.

## 5.1. Hardware

The sensor nodes used in the network are the W24TH nodes, shown in Figure 2a. The hardware core of each node is based on the NXP JN5148 module, which is an ultra-low-power, high-performance wireless microcontroller targeted at ZigBee PRO networking applications. The device features an enhanced 32-bit RISC processor, a 2.4-GHz IEEE 802.15.4 compliant transceiver, 128 kB of ROM, 128 kB of RAM and several analogue and digital peripherals. Its power consumption is 15 mA for transmission and 18 mA for reception, which gives 35% of power saving when compared to other similar platforms (e.g., TelosB).

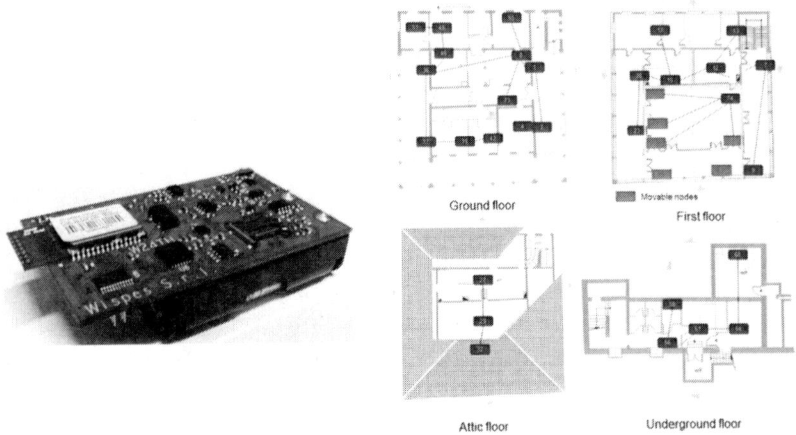


(a) (b)

**Figure 2. (a)** The W24TH sensor node and **(b)** the deployment map.

The board is equipped with several sensors such as a temperature, humidity, ambient light and a metal-oxide gas sensor. Each sensor is equipped with a specific conditioning circuit, which is useful for various ambient monitoring applications. A MicroSD card reader is also available, intended for local back-up, data logging and firmware updates. Moreover, each node is equipped with a USB battery charger, a 32-kHz quartz oscillator and an expansion connector. The power management subsystem provides the ability to switch off the whole system by software, thus enabling a 10 μA sleep mode current consumption. In addition, all sensors can be individually disconnected for the efficient management of the hardware resources, reducing the node's power consumption.

For our installation, a network of 23 low-power sensor nodes is deployed across the three floors of the historic building, as shown in Figure 2b. The topology is a tree network with a central collecting sink that is in charge of gathering the data from the nodes and forwarding them to a remote PC, used to log the data and reconstruct the missing samples.

## 5.2. Network

Generally, in low-power sensor nodes, the communication subsystem has a much higher energy consumption than the computation and sensing subsystems; hence, the radio management and duty cycling are crucial aspects for energy efficiency. Ideally, we could turn on the radio as soon as a new data packet becomes available for sending and switch it off again when the data are sent. The problem with this technique is that in collaborative networks, such as WSNs, coordination among nodes is required to enable data exchanging and forwarding from the source node to the sink. Thus, a sleep/wakeup scheduling algorithm is required to permit the correct operation of the network.

To solve this problem, we developed and implemented a new power management and scheduling protocol, named conservative power scheduling (CPS). This protocol is built on top of the IEEE 802.15.4 MAC protocol and adapts a slotted approach derived from time division multiple access (TDMA) schemes for the staggered wakeup pattern [42].

CPS is based on a centralized approach and provides a time synchronization by using a non-beacon-enabled mode and the dissemination of a service-packet. Moreover, the algorithm avoids collisions during the data packet transmission through a TDMA technique, where each node in the network transmits its data in a given time slot. The service-packet contains several pieces of information, including data for time synchronization, as well as the sequence of active sensors for all receivers. The transmission of the service-packet across the network is performed by using a novel approach based on a constructive interference of IEEE 802.15.4 symbols for fast network flooding and implicit time synchronization [43].

In this protocol, each communication period, identified by a service-packet, is divided into two parts: an active interval (AI), during which the node must keep its radio on to exchange packets with other nodes, and a sleep interval (SI), during which the node is sleeping. To permit the correct data exchange through the network, the talk interval between a node and its parent/child must overlap for at least a time slot, needed to exchange the desired data.

The real challenge of the CPS protocol is to correctly assign the time slots to each node, so as to minimize the power consumption. To achieve this goal, each node has to be in the active interval for the shortest time possible, waking up just in time for sending sampled data to the parent node and going back to sleep as soon as it has forwarded all of the data coming from its child nodes. Different from other approaches found in the literature [44], CPS does not calculate at run-time the slot scheduling, but it uses a static scheduler computed off-line. The algorithm for slot assignment takes as input the tree-structure of the network and gives as output the slot times assigned to each node.

The algorithm is articulated in two different steps: (1) ordering and (2) scheduling. In the ordering phase, each node in the network is assigned a unique ID that defines the priority of the node, following a depth-first approach. After each node is opportunely tagged with an ID, the scheduling step of the algorithm is in charge of computing the start and duration of the active time slot for each node. It ensures that each node waits until data sent by its own parent have reached the sink before sending out its own data, avoiding collisions in the network. Afterwards, the node can switch to sleep mode only after having forwarded data coming from all of its child nodes. An example of the protocol's scheduling is illustrated in Figure 3.

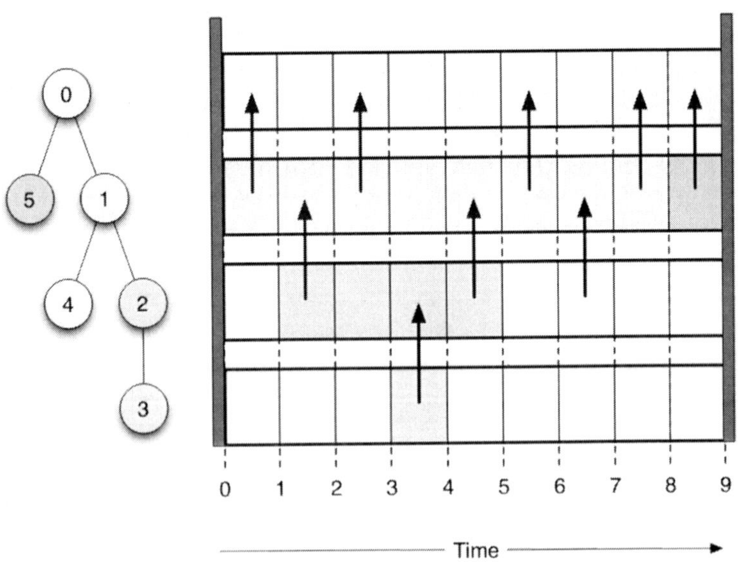


**Figure 3.** Example of scheduling using the conservative power scheduling (CPS) protocol.

In a practical implementation of CPS, the synchronization using broadcast packets from the coordinator can be problematic due to the delay in the packet relay through the network. For this reason, the margin for synchronization is relaxed, accounting for clock drift and delays in the transmission of synchronization packets.

## 5.3. Power Model

We introduce an architecture-level power model to evaluate the energy consumption of the node when the sub-sampling parameters are changed. Using this power model, fitted and validated with data from real hardware and measurements, we can easily evaluate how changing the parameters influences the energy consumption and the lifetime of the network.

Starting from the assumption reported in the previous section, the average energy consumption in each period of duration $T_k$, for a sub-sampling factor $\rho$ applied to a block of $N$ sampling intervals, is:

$$E_k = \rho\left(E_{\text{setup}} + E_{\text{sampl}} + E_{\text{store}}\right) + E_{\text{sleep}} + N^{-1}\left(E_{\text{nv}} + E_{\text{send}}\right) \quad (8)$$

where $E_{\text{sleep}}$ is the energy spent in sleep mode, $E_{\text{setup}}$ is the energy used for waking up and setting up the device, $E_{\text{Sample}}$ is the energy for sampling each sensors, $E_{\text{send}}$ is the energy used to send the acquired data, $E_{\text{store}}$ is the energy to store the acquired sample in non-volatile memory and $E_{\text{nv}}$ is the energy spent to read the data from non-volatile memory.

Expanding each term, we have:

$$\begin{aligned}E_k = \rho(&T_{\text{setup}} \cdot (P_{\text{mcu}} + P_{\text{soff}} + P_{\text{toff}}) + T_{\text{sample}} \cdot (P_{\text{sample}} + P_{\text{sactive}} + P_{\text{toff}}) + \\ &T_{\text{store}} \cdot (P_{\text{soff}} + P_{\text{toff}} + P_{\text{store}})) + T_{\text{sleep}} \cdot (P_{\text{sleep}} + P_{\text{soff}} + P_{\text{toff}}) + \\ N^{-1}(&T_{\text{nv}} \cdot (P_{\text{store}} + P_{\text{soff}} + P_{\text{toff}}) + T_{\text{send}} \cdot (P_{\text{comm}} + P_{\text{soff}} + P_{\text{send}}))\end{aligned} \quad (9)$$

where $T_{\text{sleep}}$, $T_{\text{setup}}$, $T_{\text{sample}}$, $T_{\text{send}}$, $T_{\text{store}}$, $T_{\text{nv}}$ are the duration of each respective phase, $P_{\text{sleep}}$ is the power consumed in sleep mode, $P_{\text{soff}}$ is the power absorbed from sensors when sleeping, $P_{\text{toff}}$ is the power consumption of the transceiver when the node is in sleep mode, $P_{\text{mcu}}$ is the power consumed by the MCU, $P_{\text{sample}}$ is the power spent for data acquisition, $P_{\text{sactive}}$is the power consumed by sensors, $P_{\text{comm}}$ is the power consumption for filling the transceiver output buffer and, finally, $P_{send}$ is the power for sending data. All of the values for the power consumption and timings are obtained by the values reported in Table 1 measured on real hardware or extracted from the datasheets.

**Table 1.** Characteristics of the device taken as a reference in the power model.

| $T_{sample}$ | **600 (s)** | **Sampling Period** |
|---|---|---|
| $T_{setup}$ | 5e–4 (s) | Setup time |
| $T_{store}$ (16 bit) | 5.25e–5 (s) | Time to store data in NVM |
| $V_{batt}$ | 3.3 (V) | Battery voltage |
| $A_{sleep}$ | 1.3e–6 (A) | Sleep current |
| $A_{mcu}$ | 7.5e–3 (A) | Idle current |
| $A_{sample}$ | 1.1e–3 (A) | ADC current |
| $A_{store}$ | 7.5e–3 (A) | Current when saving in NVM |
| $A_{comm}$ | 1e–6 (A) | Current for filling the transceiver buffer |
| $F$ | 24 (MHz) | Microcontroller frequency |
| $A_{send}$ | 3.1e–2 (A) | Transmission current |
| $A_{toff}$ | 10e–6 (A) | Sleep current of the transceiver |
| $S_{tx}$ | 150 (kbps) | Transmission throughput |
| $T_{setup\_radio}$ | 5e–3 (s) | Transceiver setup time |
| $B_{pkt}$ | 127 (byte) | Packet size |
| $B_{header}$ | 10 (byte) | Header size |
| $A_{sactive,TH}$ | 3e–4 (A) | Temp and Hum.(TH) current consumption |
| $A_{soff,TH}$ | 1.5e–7 (A) | TH sleep current |
| $T_{sampl,TH}$ | 2e–5 (s) | TH sampling time |
| $A_{sactive,AL}$ | 150e–6 (A) | Ambient light (AL) current consumption |
| $A_{soff,AL}$ | 0.01e–6 (A) | AL sleep current |
| $T_{sampl,AL}$ | 1e–5 (s) | AL sampling time |

# 6. RESULTS

In this section, we compare the reconstruction performance of CS and the latent variable-based statistical model against the 3ENCULT deployment, considering data coming from the temperature, humidity and light sensors. We want to investigate whether a better reconstruction technique exists among those proposed here and how the sub-sampling parameters affect such recovery quality

We can consider that each node samples the signals for a period of time $T$, called the acquisition period, ideally gathering $N = T/f_s$ samples at a $f_s$ sampling frequency The compression phase is the same for both frameworks: each node samples the signals of interest gathering a sub-set $M$ of the needed samples ($M = \rho N$), with an under-sampling ratio $0 < \rho < 1$. The under-sampling pattern is locally generated by each node using its own ID and the time-stamp as the seed for randomization. In the random sampling pattern, the inter-measurements intervals are always multiples of the minimum sampling period $T_k = T/N$.

After the acquisition period $T = NT_k$, the gathered data are sent to the collecting sink through the network. The sampling time $T_k$ in the following simulations is set to 600 s and the results are averaged over 100 trials. Each trial is characterized by a different sampling pattern and a different considered portion of the signal. The reconstruction phase is fairly different and determines the recovery quality of the original signal. For CS, the DCTmatrix is used as the sparsifying matrix, which has already been demonstrated to be a good sparsifying matrix for natural signals [45,46].

In the first simulation, we reconstruct the original signals from a sub-sampled version without exploiting any inter- or intra-correlation among them, just averaging the reconstruction quality over all of the signals, with a signal length of $N = 512$. For the latent variables approach, here, we used the standard CP tensor factorization technique, without the contribution of any correlations in the data. The comparison is carried evaluating the signal-to-noise ratio (SNR), defined as:

$$\mathrm{SNR_{dB}} = 20 \cdot \log_{10} \frac{\|\mathbf{x}\|_2}{\|\mathbf{x} - \hat{\mathbf{x}}\|_2} \tag{10}$$

where $\mathbf{x}$ is the original signal and $\hat{\mathbf{x}}$ is its recovered version. We show the average SNR across all of the network nodes.

From the plot, shown in Figure 4a, we can infer how different the performance is for the two techniques: while the reconstruction performance for LV is pretty stable when varying the sub-sampling factor $\rho$, CS is much more affected by the compression factor. Recovery with CS achieves a better reconstruction almost for every sub-sampling factor with respect to the latent variable-based technique.

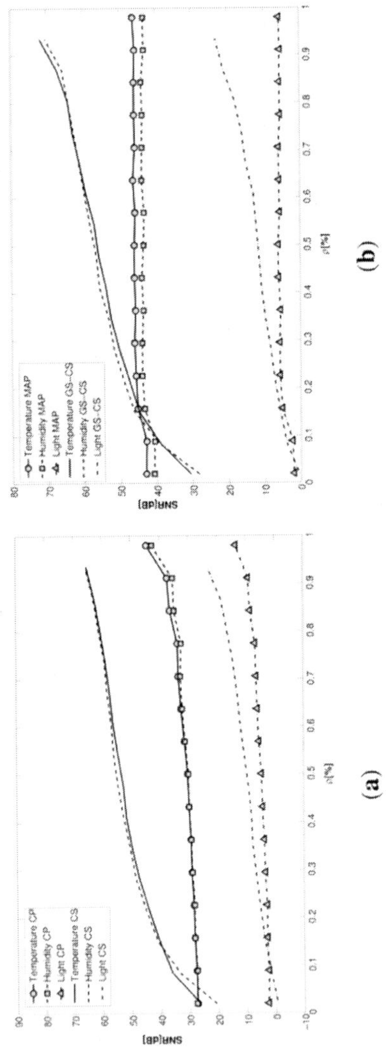


**Figure 4.** Recovery comparison when reconstructing the original signals from sub-sampled versions ($N = 512$): **(a)** compressive sensing (CS) and latent variable (LV) tensor factorization averaging the reconstruction quality over all of the nodes; **(b)** GC-CS and maximum *a posteriori* (MAP), exploiting the correlations among sensors and nodes.

Both of the frameworks seem to be greatly affected by the nature of the signal to reconstruct. In particular, from the plot, we can infer how difficult the recovery of light signals is for the two proposed techniques. This is due to the nature of the light signal that is recorded inside the building. While for temperature and humidity, the gathered signals are continuous signals and smoothly affected by human presence, the light signal is highly irregular and highly influenced by the artificial lighting in single rooms. Moreover, some of the nodes are placed in the basement, where the light level is under the noise threshold of the light sensors, providing extremely noisy data.

To evaluate whether it is possible to exploit the correlations existing among sensors and nodes to improve the reconstruction, we performed the recovery against the same dataset using for CS the group sparse optimization (GS-CS), exploiting the joint sparsity of the signals, and for LV the maximum *a posteriori* optimization (LV-MAP) introduced inEquation (8). The results are reported in Figure 4b. While the performance for CS remains almost the same, the LV-MAP method guarantees a significant improvement in reconstruction, resulting in better performance than CS for small values of $\rho$, especially in relation to humidity and temperature signals.

The behavior of CS could be explained by looking at Figure 5. Here, we show how the union over all signals of the $K$ best DCT basis vectors per signal has a size definitely greater than $K$. Practically, this means that GS-CS is able to exploit the inter-node correlation only to a small extent, since the shared information among different nodes is limited, and the recovery algorithm is not able to exploit this information to improve the recovery quality.

According to the model in Section 5.2, the simulations are performed with a sampling frequency $f_s = 1/600$ Hz, and since the length of the data is $N = 512$, this brings a delay in the data delivery towards the data collector of 3.5 days. Thus, the size of the recovered signal spans across 3.5 days. Having high values of $N$ means that we have to wait a longer time to proceed with data recovery. Therefore, we want now to investigate how the length of the block of data gathered by the sensors affects the two frameworks and whether a correlation between recovery performance and the $N$ parameter exists for GS-CS and LV-MAP.

In Figure 6a, the results for GS-CS when $N$ is changed are presented. From the plot, we can infer how the length of the signal $N$ does not greatly affect the reconstruction quality for all of the signals taken into consideration. Rather, we can see how the influence of the parameter $N$ (and then, of the delay in the data collection) is only visible for small values of the sub-sampling factor ρ. Different from temperature and humidity, the light signal presents a peculiar behavior showing an increased reconstruction quality with the increase in the number of acquired samples.

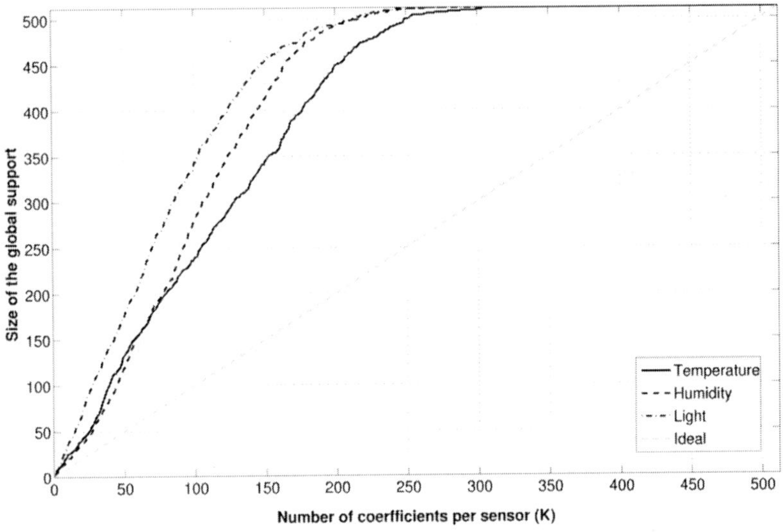


**Figure 5.** Number of DCTcoefficients necessary to include the $K$ largest coefficients for each signal ($N = 512$).

The same results for the LV-MAP approach are presented in Figure 6b. The difference in the reconstruction error for the various values of $N$ is more evident than in the GS-CS case. With small values of $N$, we registered difficulties in reconstructing the desired signals. The best recovery performance is achieved when considering 256–512 samples at a time, identifying the optimal trade-off between delay and reconstruction accuracy, since larger blocks of data present again a loss of accuracy. For a delay smaller than $N = 128$, the reconstruction of the light signal is not feasible in both cases, since the majority of the samples gathered are zeros, due to the lack of light at night.

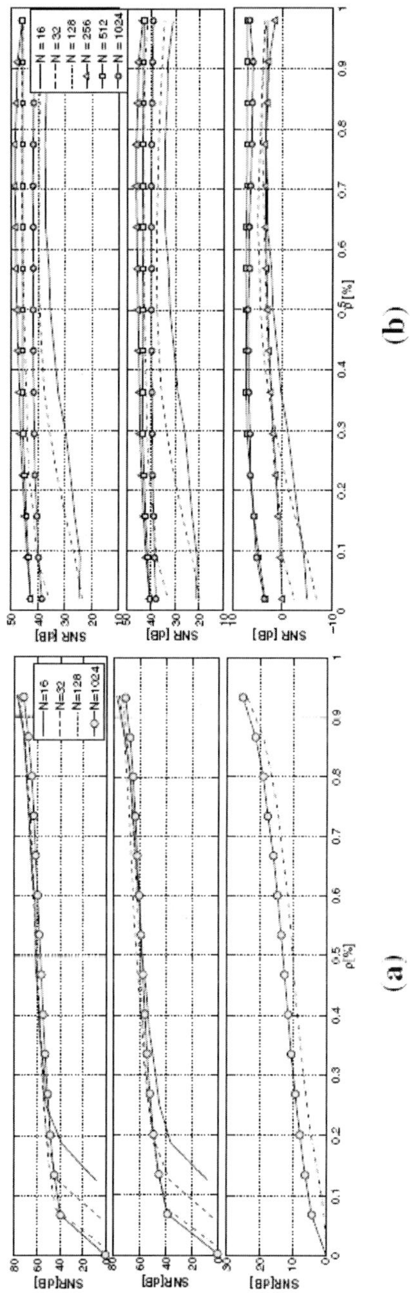

**Figure 6.** Reconstruction quality varying the sub-sampling factor ρ and using the signal length *N* as the parameter: **(a)** group sparse (GS)-CS and **(b)** LV-MAR From top to bottom: temperature, humidity, light.

Having evaluated the recovery performance and the influence of the gathering delay on reconstruction, it is interesting to investigate the power consumption involved with compression according to the power model in Section 5.3. In Figure 7, we report the reconstruction quality against the energy consumption for one acquisition cycle. The plot clearly shows how a trade-off between energy consumption and transmission delay does exist in the GS-CS case (Figure 7a). Higher values of $N$, thus higher delays in transmission, are able to guarantee a better reconstruction quality with definitely less energy than the$N = 16$ case (all of the other cases are not considered in the plot, since they are between these two boundaries). The light signal is a special case, but we can draw the same conclusions as before. The LV-MAP case (Figure 7b) presents a similar behavior, but with a less emphasized increase in energy efficiency corresponding to the increase in data size $N$. When comparing the two graphs, we can observe how the GS-CS case exhibits a slightly higher energy efficiency, allowing a higher reconstruction quality when considering the same energy consumptions as the LV-MAP case. Only for extremely sub-sampled signals are the LV-MAP approach's results better, having a major benefit from the explicit correlation models incorporated in the data reconstruction. In both cases, we are able to obtain a better accuracy (or the same reconstruction quality with less energy) if we are willing to wait for a higher number of gathered samples before proceeding with the reconstruction.

The same conclusions are shown in Figure 8a, where the ratio between the reconstruction quality and the consumed energy is plotted against the sub-sampling factor $\rho$. Here, we can see a direct comparison of the two techniques for the case with the best reconstruction performance ($N = 512$). Again, we can see how the GS-CS case has a higher ratio when compared to the LV-MAP case, for almost all of the sub-sampling policies. Only when dealing with a really small amount of sampled data (20%) does the LV-MAP case show a better performance. This result can be a guide for WSN developers, suggesting the adoption of the LV method only when a really aggressive power saving technique is needed. The plots of the two techniques are combined in Figure 8b, where the best ratio between the reconstruction accuracy and the consumed energy is plotted against the sub-sampling factor. Here, we have the combination of the two approaches and show the best achievable results for every sub-sampling factor.

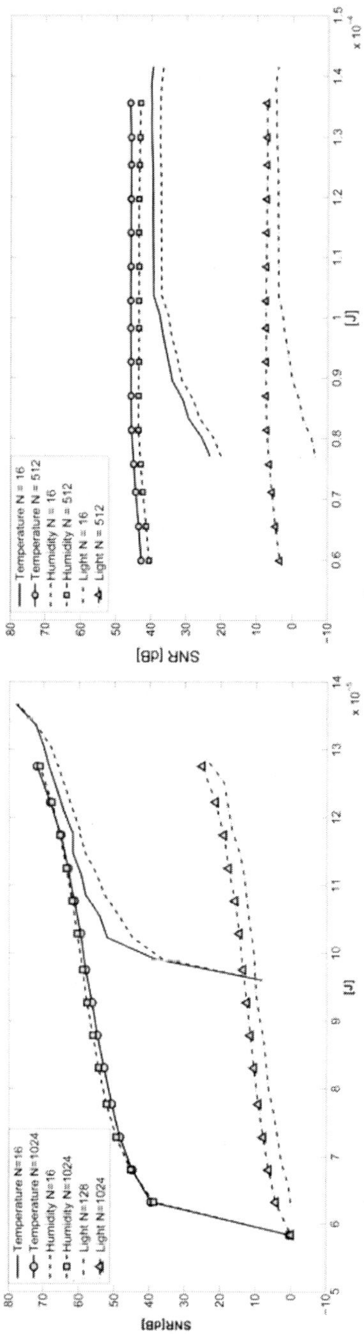


**Figure 7.** Reconstruction quality *vs.* averaged per cycle energy consumption varying the parameter *N*: **(a)** CS and **(b)** LV.

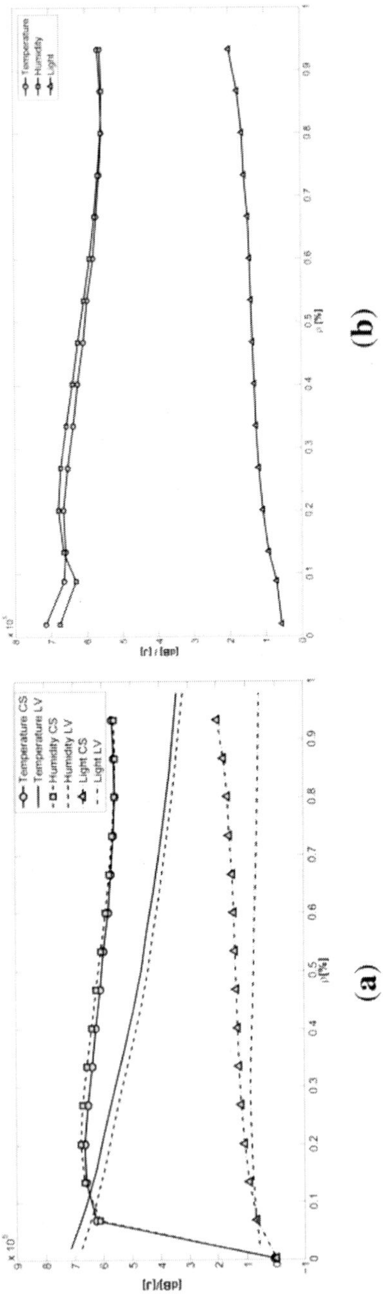


**Figure 8. (a)** Ratio between the recovery quality and energy spent in compression varying the sub-sampling factor ρ for the two approaches; **(b)** the combination of the two approaches.

There is no detailed analysis in the literature on conjunct sub-sampled data reconstruction and accurate power consumption estimation for heterogeneous sensor networks. In [28], an SNR of 40 to 68 dB is reported for the reconstruction of temperature signals from compressed data gathered at sub-sampling factors ranging from 0.1 to 0.8, but no power analysis is performed. The hybrid CS aggregation scheme of [18] achieves 20% to 50% energy savings for different aggregation factors, when compared to non-aggregation techniques on a simulated 1255-node grid network. The paper, however, does not report any detail on the power consumption models used, nor on the impacts of the proposed CS aggregation algorithm on the reconstructed data quality. In this work, we presented two energy-efficient approaches, with detailed data reconstruction and power consumption analysis on a heterogeneous WSN, introducing the use of spatio-temporal correlations to improve the performance. Considering the temperature signal, we achieve 40 to 70 dB of SNR for increasing sub-sampling factors, which corresponds to average power consumptions ranging from 63 to 128 μJ, with maximum energy savings of 58%. Our results are aligned with the presented literature and introduce detailed multi-signal reconstruction and power analysis based on a real-life WSN deployment, which provide insights into the application of sub-sampling in heterogeneous WSNs and can serve as guidelines and future design specifications.

# 7. CONCLUSIONS

In this paper, we presented a wireless sensor network deployment and used it to compare two promising techniques for energy-efficient data gathering and reconstruction. One is based on the well-known compressive sensing framework, while the second is a latent variable-based statistical model. Both approaches try to successfully recover the desired signal from a highly incomplete sub-sampled version, obtained opportunely, skipping samples during the acquisition phase. The two techniques exploit redundancies and correlations present in the gathered data to achieve a better reconstruction accuracy with a smaller number of collected samples and, thus, with a lower energy consumption. We described the hardware and software implementation of the deployed network, introduced an energy model for the sensor nodes and analyzed the gathered data, comparing the two reconstruction techniques. The results showed how the use of the data from the whole network and their correlations improved the recovery performance in both cases, when compared to the standard approaches, where individual nodes and signals are considered. The CS approach usually achieves better reconstruction accuracy, with the exception of cases when really aggressive sub-sampling policies are used. This leads also to the better energy efficiency of the CS method.

## ACKNOWLEDGMENTS

The research leading to these results has received funding from EU Projects Genesi (Grant Agreement No. 257916) and 3ENCULT (Grant Agreement No. 260162), funded by the EU 7th Framework Programme. The authors would like to thank Tajana Simunic Rosing and Piero Zappi for helpful discussion, valuable comments and support.

## AUTHOR CONTRIBUTIONS

All of the authors have contributed to the development and writing of this article. Bojan Milosevic and Carlo Caione carried out the hardware deployment and measurements, developed the MATLAB code for data collection and for the testing of the reconstruction algorithms and worked on the draft of the paper. Elisabetta Farella and Davide Brunelli supervised the technical work and contributed to the writing of the paper. Luca Benini supervised the whole work, corrected the draft of the paper and contributed to the writing of the final revision of the paper.

## REFERENCES

1. Li, X.Y.; Wang, Y.; Wang, Y. Complexity of Data Collection, Aggregation, and Selection for Wireless Sensor Networks.*IEEE Trans. Comput.* **2011**, *60*, 386–399.

2. Ammari, H.M. Investigating the Energy Sink-Hole Problem in Connected k-Covered Wireless Sensor Networks. *IEEE Trans. Comput.* **2013**, *63*, 2729–2742.

3. Li, M.; Li, Z.; Vasilakos, A.V. A survey on topology control in wireless sensor networks: Taxonomy, comparative study, and open issues. *IEEE Proc.* **2013**, *101*, 2538–2557.

4. Moser, C.; Thiele, L.; Brunelli, D.; Benini, L. Adaptive Power Management for Environmentally Powered Systems.*IEEE Trans. Comput.* **2010**, *59*, 478–491.

5. Zappi, P.; Lombriser, C.; Stiefmeier, T.; Farella, E.; Roggen, D.; Benini, L.; TrÃűster, G. Activity Recognition from On-Body Sensors: Accuracy-Power Trade-Off by Dynamic Sensor Selection. Proceedings of the European Conference on Wireless Sensor Networks, Bologna, Italy, 30 January–1 February 2008; pp. 17–33.

6. Veeravalli, V.V.; Varshney, P.K. Distributed inference in wireless sensor networks. *Philos. Trans. R. Soc. A Math. Phys. Eng. Sci.* **2012**, *370*, 100–117.

7. Zou, Z.; Hu, C.; Zhang, F.; Zhao, H.; Shen, S. WSNs data acquisition by combining hierarchical routing method and compressive sensing. *Sensors* **2014**, *14*, 16766–16784.

8. Aquino, A.L.; Junior, O.S.; Frery, A.C.; de Albuquerque, E.L.; Mini, R.A. MuSA: Multivariate sampling algorithm for wireless sensor networks. *IEEE Trans. Comput.* **2014**, *63*, 968–978.

9. AbdelSalam, H.S.; Olariu, S. Toward adaptive sleep schedules for balancing energy consumption in wireless sensor networks. *IEEE Trans. Comput.* **2012**, *61*, 1443–1458.

10. Candes, E.; Wakin, M. An introduction to compressive sampling. *IEEE Signal Process. Mag.* **2008**, *25*, 21–30.

11. Duarte, M.; Eldar, Y. Structured compressed sensing: From theory to applications. *IEEE Trans. Signal Process.* **2011**, *59*, 4053–4085.

12. Verma, N.; Zappi, P.; Rosing, T. Latent variables based data estimation for sensing applications. Proceedings of the IEEE International Conference on Intelligent Sensors, Sensor Networks and Information Processing, Adelaide, Australia, 6–9 December 2011; pp. 335–340.

13. Milosevic, B.; Yang, J.; Vrema, N.; Tilak, S.; Zappi, P.; Farella, E.; Rosing, T.S.; Benini, L. Efficient energy management and data

recovery in sensor networks using latent variables based tensor factorization. Proceedings of the ACM/IEEE International Conference on Modeling, Analysis and Simulation of Wireless and Mobile Systems (MsWiM), Barcelona, Spain, 3–8 November 2013; pp. 247–254.

14. Caione, C.; Brunelli, D.; Benini, L. Distributed Compressive Sampling for Lifetime Optimization in Dense Wireless Sensor Networks. *IEEE Trans. Ind. Inform.* **2012**, *8*, 30–40.

15. Mamaghanian, H.; Khaled, N.; Atienza, D.; Vandergheynst, P. Design and Exploration of Low-Power Analog to Information Conversion Based on Compressed Sensing. *IEEE J. Emerg. Sel. Top. Circuits Syst.* **2012**, *2*, 493–501.

16. Koren, Y.; Bell, R.; Volinsky, C. Matrix Factorization Techniques for Recommender Systems. *Computer* **2009**, *42*, 30–37.

17. Kolda, T.G.; Bader, B.W. Tensor Decompositions and Applications. *SIAM Rev.* **2009**, *51*, 455–500.

18. Xiang, L.; Luo, J.; Vasilakos, A. Compressed data aggregation for energy efficient wireless sensor networks. Proceedings of the Sensor, Mesh and Ad Hoc Communications and Networks (SECON), Salt Lake City, UT, USA, 27–30 June 2011; pp. 46–54.

19. Yao, Y.; Cao, Q.; Vasilakos, A. EDAL: An Energy-Efficient, Delay-Aware, and Lifetime-Balancing Data Collection Protocol for Wireless Sensor Networks. Proceedings of the 2013 IEEE 10th International Conference on Mobile Ad-Hoc and Sensor Systems (MASS), Hangzhou, China, 14–16 October 2013; pp. 182–190.

20. Baron, D.; Duarte, M.F.; Wakin, M.B.; Sarvotham, S.; Baraniuk, R.G. Distributed Compressive Sensing. Available online: arxiv.org/pdf/0901.3403 (accessed on 28 February 2015).

21. Lee, K.; Bresler, Y.; Junge, M. Subspace Methods for Joint Sparse Recovery. *IEEE Trans. Inf. Theory* **2012**, *58*, 3613–3641.

22. Duarte, M.; Baraniuk, R. Kronecker Compressive Sensing. *IEEE Trans. Image Process.* **2012**, *21*, 494–504.

23. Deng, W.; Yin, W.; Zhang, Y. Group Sparse Optimization by Alternating Direction Method. Rice CAAM Report TR11-06.

Available online: http://www.caam.rice.edu/optimization/L1/GroupSparsity/group110419.pdf (accessed on 28 February 2015).

24. Su, X.; Khoshgoftaar, T.M. A survey of collaborative filtering techniques. *Adv. Artif. Intell.* **2009**.

25. Yoshii, K.; Goto, M.; Komatani, K.; Ogata, T.; Okuno, H. An Efficient Hybrid Music Recommender System Using an Incrementally Trainable Probabilistic Generative Model. *IEEE Trans. Audio Speech Lang. Process.* **2008**, *16*, 435–447.

26. Koren, Y. Collaborative filtering with temporal dynamics. *Commun. ACM* **2010**, *53*, 89–97.

27. Karatzoglou, A.; Amatriain, X.; Baltrunas, L.; Oliver, N. Multiverse recommendation: N-dimensional tensor factorization for context-aware collaborative filtering. Proceedings of the Fourth ACM Conference on Recommender Systems, Barcelona, Spain, 26–30 September 2010; pp. 79–86.

28. Luo, C.; Wu, F.; Sun, J.; Chen, C.W. Efficient Measurement Generation and Pervasive Sparsity for Compressive Data Gathering. *IEEE Trans. Wirel. Commun.* **2010**, *9*, 3728–3738.

29. Mahmudimanesh, M.; Khelil, A.; Suri, N. Reordering for Better Compressibility: Efficient Spatial Sampling in Wireless Sensor Networks. Proceedings of the 2010 IEEE International Conference on Sensor Networks, Ubiquitous, and Trustworthy Computing, Newport Beach, CA, USA, 7–9 June 2010; pp. 50–57.

30. Mamaghanian, H.; Khaled, N.; Atienza, D.; Vandergheynst, P. Compressed Sensing for Real-Time Energy-Efficient ECG Compression on Wireless Body Sensor Nodes. *IEEE Trans. Biomed. Eng.* **2011**, *58*, 2456–2466.

31. Makhzani, A.; Valaee, S. Reconstruction of a Generalized Joint Sparsity Model Using Principal Component Analysis. Proceedings of the 2011 4th IEEE International Workshop on Computational Advances in Multi-Sensor Adaptive Processing (CAMSAP), San Juan, Puerto Rico, 13–16 December 2011; pp. 269–272.

32. Murray, T.S.; Pouliquen, P.O.; Andreou, A.G. Design of a Parallel Sampling Encoder for Analog to Information (A2I) Converters:

Theory, Architecture and CMOS Implementation. *Electronics* **2013**, *2*, 57–79.

33. Yu, Z.; Hoyos, S. Digitally Assisted Analog Compressive Sensing. Proceedings of the 2009 IEEE Dallas Circuits and Systems Workshop, Richardson, TX, USA, 4–5 October 2009; pp. 1–4.

34. Baheti, P.; Garudadri, H. An Ultra Low Power Pulse Oximeter Sensor Based on Compressed Sensing. Proceedings of the Sixth International Workshop on Wearable and Implantable Body Sensor Networks, Berkeley, CA, USA, 3–5 June 2009; pp. 144–148.

35. Garudadri, H.; Baheti, P.; Majumdar, S.; Lauer, C.; Massé, F.; van de Molengraft, J.; Penders, J. Artifacts mitigation in ambulatory ECG telemetry. Proceedings of the 2010 12th IEEE International Conference on e-Health Networking Applications and Services (Healthcom), Lyon, France, 1–3 July 2010; pp. 338–344.

36. Prada, M.A.; Toivola, J.; Kullaa, J.; Hollmen, J. Three-way analysis of structural health monitoring data.*Neurocomputing* **2012**, *80*, 119–128.

37. Chen, W.; Wassell, I. Energy efficient signal acquisition via compressive sensing in wireless sensor networks. Proceedings of the 2011 6th International Symposium on Wireless and Pervasive Computing (ISWPC), Hong Kong, China, 23–25 February 2011; pp. 1–6.

38. Needell, D.; Vershynin, R. Uniform Uncertainty Principle and Signal Recovery via Regularized Orthogonal Matching Pursuit. *Found. Comput. Math.* **2009**, *9*, 317–334.

39. Chen, W.; Rodrigues, M.; Wassell, I. Distributed compressive sensing reconstruction via common support discovery. Proceedings of the 2011 IEEE International Conference on Communications (ICC), Kyoto, Japan, 5–9 June 2011; pp. 1–5.

40. Acar, E.; Yener, B. Unsupervised Multiway Data Analysis: A Literature Survey. *IEEE Trans. Knowl. Data Eng.* **2009**, *21*, 6–20.

41. Aderohunmu, F.A.; Balsamo, D.; Paci, G.; Brunelli, D. Long Term WSN Monitoring for Energy Efficiency in EU Cultural Heritage Buildings. *Lect. Notes Electr. Eng.* **2014**, *281*, 253–261.

42. Keshavarzian, A.; Lee, H.; Venkatraman, L. Wakeup scheduling in wireless sensor networks. Proceedings of the 7th ACM International Symposium on Mobile Ad Hoc Networking and Computing, Florence, Italy, 22–25 May 2006; pp. 322–333.

43. Ferrari, F.; Zimmerling, M.; Thiele, L.; Saukh, O. Efficient network flooding and time synchronization with Glossy. Proceedings of the 2011 10th International Conference on Information Processing in Sensor Networks (IPSN); pp. 73–84.

44. Hohlt, B.; Doherty, L.; Brewer, E. Flexible power scheduling for sensor networks. Proceedings of the 3rd International Symposium on Information Processing in Sensor Networks, Berkeley, CA, USA, 26–27 April 2004; pp. 205–214.

45. Caione, C.; Brunelli, D.; Benini, L. Compressive Sensing Optimization for Signal Ensembles in WSNs. *IEEE Trans. Ind. Inform.* **2014**, *10*, 382–392.

46. Caione, C.; Brunelli, D.; Benini, L. Compressive sensing optimization over ZigBee networks. Proceedings of the 2010 International Symposium on Industrial Embedded Systems (SIES), Trento, Italy, 7–9 July 2010; pp. 36–44.

4

# AN IEEE 802.15.4 BASED ADAPTIVE COMMUNICATION PROTOCOL IN WIRELESS SENSOR NETWORK: APPLICATION TO MONITORING THE ELDERLY AT HOME

**Juan Lu[1,2], Adrien Van Den Bossche[1,3], Eric Campo[1,2]**

[1]Univ de Toulouse, Toulouse, France

[2]CNRS, LAAS, Toulouse, France

[3]CNRS, IRIT, Toulouse, France

## ABSTRACT

Monitoring behaviour of the elderly and the disabled living alone has become a major public health problem in our modern societies. Among the various scientific aspects involved in the home monitoring field, we are interested in the study and the proposal of a solution allowing distributed sensor nodes to communicate with each other in an optimal way adapted to the specific application constraints. More precisely, we want to build a wireless network that consists of several short range sensor nodes exchanging data between them according to a communication protocol at MAC (Medium Access Control) level. This protocol must be able to optimize energy consumption, trans- mission time and loss of information. To achieve this objective, we have analyzed the advantages and the limitations of WSN (Wireless Sensor Network) technologies and communication protocols currently used in relation to the requirements of our application. Then we proposed a deterministic, adaptive and energy saving medium access method based on the IEEE 802.15.4 physical layer and a mesh topology. It ensures the message delivery time with strongly limited collision risk due to the spatial reuse of medium in the two-hop neighbourhood. This proposal was characterized by modelling and simulation using OPNET network simulator. Finally we implemented the proposed mechanisms on hardware devices and deployed a sensors network in real situation to verify the accuracy of the model and evaluate the proposal according to different test configurations.

## KEYWORDS

Wireless Sensor Network, Medium Access Control, Quality of Service, Energy Saving, Mesh Topology, IEEE 802.15.4, Indoor Monitoring Application

## 1. INTRODUCTION

Nowadays the aging population is constantly increasing and many countries are now facing an urgent requirement to provide appropriate home environment solutions for their citizens [1] [2]. These individuals attach a great importance to the autonomy that allows them to live mostly at home, and in their immediate environment, providing them freedom and a better quality of life. But, in the case of an accident such as a fall, faintness..., that autonomy can quickly turn into dependence. To supply solutions, some people use systems embedded on their body, such as physiological sensors or fall sensors [3] - [5]. These devices are intrusive and limitations become apparent due to the fact that the patient is often unable to use an alert system because either he is not wearing his equipment or, if he suddenly feels unwell, is unable to perform the alert activation gesture.

The solution we consider is to instrument the environment of the person. Indeed, by monitoring the main environmental characteristics of their living space, it seems to be possible to get a lifestyle pattern of the person [6] -[8]. For example, measuring temperature, humidity, luminosity, noise levels, presence..., in many strategic areas at home can provide useful data to interpret a physical activity in space and time.

Data processing will determine circadian activity rhythms of the person and so will contribute to detect unusual situations and emergency cases. Generally, the challenge is to propose a suitable sensor network that allows uninterrupted data transmission in a bounded time.

Within this context, the objective of this work is to modelize and implement a complete heterogeneous sensor network allowing the measurement and the transmission of short-range data collected by the environmental sensors. The planned network will be deployed in a house or even building and transmit alert messages caused by a malfunction of environmental parameters via a continuous monitoring. So a limited scale, up to 50 nodes, seems to be sufficient for a home monitoring application. These nodes exchange data between them according to a communication protocol and, obviously, the following performance criteria are crucial to an indoor monitoring application: 1) Different QoS capacities should be provided for message delivery, especially when some time- sensitive messages are zero tolerance for packet loss. 2) Power consumption is a fundamental concern for the battery constrained sensor nodes. A long lifetime network, several months or even years, is eagerly expected. 3) The network should be self-organizing, quickly deployed and robust against link failure or link establishment. For example, the failure of critical nodes can lead to the entire network failure and harm the safety of people monitored by this network.

Our work focuses on MAC layer to address the above constraints. A mesh WSN is considered because it is more flexible and robust than a star or a tree topology. However, how to construct this topology and how to provide the QoS are some issues in a mesh WSN. The avoidance of collisions between 2-hop neighbours is another goal because there is scarcely interference at distance of more than 2 hops [9] . Also, as the largest energy consumption of the nodes is due to the time spent in the receive state [10] , so time slot allocation in a mesh topology is an important task to strictly limit non efficient transceiver activities such as idle listening.

This paper aims to present a beacon-enabled MAC over IEEE 802.15.4 PHY which supports mesh topology and enables guaranteed service with low energy consumption. The paper is organized as follows: in Section 2 some related works are presented. In Section 3 the proposed MAC protocol is fully described. Section 4 provides simulation results and Section 5 presents the results obtained through practical measurements in a real environment. The last section concludes the paper.

# 2. RELATED WORKS

Several standards for low-power, low-cost and short-range network have been investigated. Among them, IEEE 802.15.4 is considered as a promising way in terms of energy saving and guaranteed medium access. Many other main standards such as ZigBee, IEEE 802.15.5 and 6LoWPAN are all based on IEEE 802.15.4 PHY/MAC or backwards compatible with this standard. Therefore, we consider IEEE 802.15.4 as a starting point for our work. In part 2.1, IEEE 802.15.4 standard is briefly presented. In part 2.2, previous scientific works including classical WSN MAC protocols and recent works based on IEEE 802.15.4, have been studied and analyzed. The focus is mainly on the metrics: QoS, energy saving and mesh topology.

## 2.1. IEEE 802.15.4 Standard

IEEE 802.15.4 [11] is a well-known standard for Low-Power and Low-Rate Wireless Personal Area Networks (LP-LR-WLAN). Several physical layers (PHYs) are proposed; the main and widely used 802.15.4 PHY operates in the 2.4 GHz radio band and enables transmissions at 250 kbps in a 30 mindoor range. The protocol supports beacon and non beacon mode. More precisely, in beacon mode, it is possible to achieve variable duty cycles (from 100% down to 0.006%), which is particularly interesting for our application where energy constraint and network lifetime are concerned. In addition, the beacon mode has an attractive feature for time-sensitive applications as QoS properties are available with GTS (Guaranteed Time Slot) mechanism. On the other side, the non beacon mode that has the advantage of a lower complexity and a higher scalability as compared with beacon mode, does not provide any of those features. Thus, we focus our development on the beacon mode of this standard.

IEEE 802.15.4 may operate in two topologies: a star topology or a peer-to-peer topology. Peer-to-peer topology allows a more complex network to be implemented such as cluster-tree topology. However, how to construct a mesh networking topology as required by our application is missing in the standard. In fact, while the current standard supports multi-hop networking using peer-to-peer topology, it restricts its use to a non beacon mode. This contradiction makes the interesting advantages, such as GTS and energy saving, disappear, especially on router nodes.

## 2.2. MAC Protocols for WSN

Some classical WSN MAC protocols such as S-MAC [12] and ContikiMAC [13] and their optimized works such as T-MAC [14] and X-MAC [15] are based on the sleep-wake cycle to reduce energy consumption of WSN. However, they all access the medium by CSMA (Carrier Sense Multiple Access) mechanism that provides a best-effort service. So the transmission of time-sensitive messages may not be guaranteed.

ZigBee specifications [16] clear the ambiguities of IEEE 802. 15.4 in a cluster-tree topology. The centralized PAN (Personal Area Network) coordinator calculates and assigns a beacon transmission offset for each node when it wants to associate the PAN. Therefore, the network scalability and flexibility are both limited. The direct communications between neighbours are not possible.

Anis Koubâa also focuses his work in the field of cluster-tree topology. In TDBS [17] [18] , the requirement of different BI (Beacon Interval) and SD (Superframe Duration) for each node is calculated in advance. However, this weakens the flexibility and robustness as well as restricts the scalability of network.

Another example has been proposed in OCARI project [19] [20] . A PAN coordinator is the destination of all association requests and allows a beacon slot for each associated node. The main drawback of this solution is the lack of flexibility, especially regarding a changing topology, for example due to the inconstancy of wireless medium.

P. S. Muthukumaran proposed MeshMAC protocol [21] . This protocol enables mesh networking over the beacon mode through a distributed SDS (Superframe Duration Scheduling) strategy in which each node calculates its schedule to transmit beacons based only on locally available information. The limitations of MeshMAC are: it imposes very low duty cycles for a large scale network so that the end-to-end delay may be extended; the beacon transmission offset is difficult to choose if the topology changes.

B. Carballido Villaverde proposed DBOP MAC protocol [22] [23] . It creates a BOP (Beacon Only Period) where beacons are transmitted at different time slots among neighbours and neighbours' neighbours. However, DBOP introduces an overhead into the IEEE 802.15.4 MAC standard. Another drawback is the inefficient management of BOP length. In addition, our MAC protocol was proposed before this work and we have continued to work on it until a prototype has been totally implemented.

Hence, we consider that all the current standards and protocols cannot satisfy all our application criteria. Many problems such as beacon collision, dynamic timeslot allocation, energy saving on router nodes, determinism medium access in a mesh topology, etc., have not been solved yet. This makes urgent requirement of new MAC protocols.

# 3. ADCF MAC PROTOCOL

This part presents an original MAC protocol named as ADCF (Adaptive and Distributed Collision Free). The proposed protocol aims to improve robustness and flexibility of IEEE 802.15.4 MAC, which means the capacity of self-organization and auto-reparation. Simultaneously, ADCF should enable energy efficiency and guaranteed slots negotiation [24] [25] . Before showing the details, some assumptions should be highlighted: 1) the main contribution of ADCF concerns the router nodes. We have considered an ad-hoc approach: all the considered nodes have the capacity to be both sensor and router, even if classical sensors (Reduced Function Devices) may be associated to the routers, but this last point is out of the scope of this paper. 2) To ensure full determinism in the build of the topology, nodes addresses must be preliminary set during installation.

## 3.1. Overview of ADCF

ADCF is based on the IEEE 802.15.4 2.4 GHz DSSS physical layer and classical superframe structure. On the one hand, ADCF proposes a distributed beacon scheduling mechanism in a beacon only period which spatially reuses the timeslots over 2-hop. On the other hand, the contribution of ADCF lies in a data slot allocation mechanism, which makes GTS possible in a mesh topology.

Generally, the nodes can switch between two stages: initialization stage and working stage. In initialization stage, which is not using the superframe structure, the nodes access medium by unslotted CSMA/CA mechanism. In working stage, the time is divided into superframe and each superframe includes three parts: {BOP, Active Period, Inactive Period}.

As shown in Figure 1, BOP is organized by CFBS (Collision Free Beacon Slot). Each node has a 2-hop collision-free beacon slot in BOP. Active period is divided into 16 equally sized slots as a classical IEEE 802.15.4 superframe. It starts with the CAP (Contention Access Period) where medium accesses are done by using slotted CSMA/CA mechanism. It ends with CFP (Contention Free Period) where medium accesses are done by using an original mechanism that provides GTS-equivalent for the mesh topology. This original mechanism is based on the CFDS (Collision Free Data Slot). After active period, the optional inactive period allows all the nodes to go to sleep mode for energy saving.

In a mesh network, the node that uses the first CFBS of the superframe is called as initiator. Initiator is not a supernode as each node in the network may be selected as initiator; initiator is not a supernode as it only has a partial knowledge of the 2-hop neighbourhood, like the other nodes; initiator is not a supernode as the network could work properly when initiator fails; initiator is not a supernode as it has not any central role such as a classical IEEE 802.15.4 PAN-Coordinator.

In the rest of this part, CFBS, CFDS mechanisms as well as a smart repair protocol will be presented.

## 3.2. CFBS Mechanism

The beacon frame is very important as it announces the presence of the node and the presence of its 1-hop neighbours; it maintains network synchronization and is used to request/reply neighbour data slot negotiations, i.e. CFDS. The nodes far away more than 2 hops could reuse the same CFBS to enhance channel reutilization. In other words, nodes must determine their CFBSs by taking into account CFBSs used by their 2-hop neighbourhood with a certain priority.

To determine priority, three parameters of the node are employed: 1) ND (Neighbour Density) indicates the number of neighbours within 2 hops, including the node itself. The node with the higher ND is usually in the center zone of a network and has a faster awareness to the topological changes. 2) NE (Neighbour Energy) indicates the residual energy level of the node and has the profound effect to the network. 3) NA (Neighbour Address) means the 16-bit address of the node and is unique for each node. The node with higher ND, higher NE and smaller NA has higher priority. The comparison order is ND, NE and then NA.

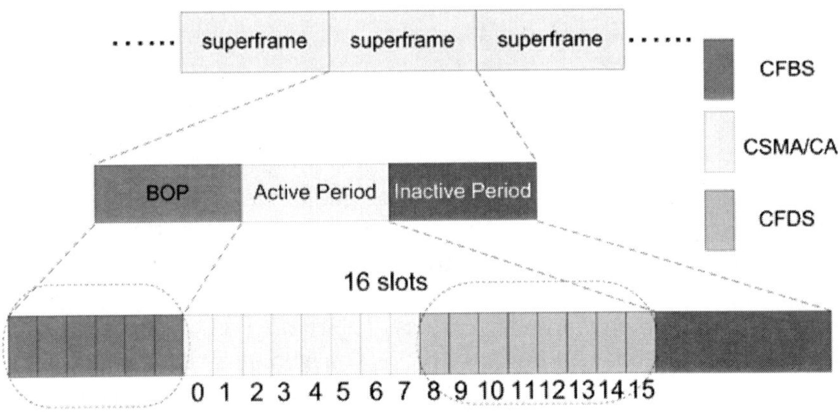


**Figure 1**. ADCF superframe structure.

The node with the highest priority in the network is selected as the initiator. It specifies the beginning of BOP and measures the length of BOP (BOPL). The initiator's ND is defined as $D_{max}$ and equals the length of BOP.

When a new node joins the network, it will firstly listen to the channel for a fixed period. Depending on the received beacons during listening, the new node will send its own beacon by different mechanisms. Each node broadcasts its beacon within 1-hop and records direct neighbours' in its NT (Neighbour Table). Therefore, all the 2-hop neighbours' information is obtained by this new node. The interesting information in a beacon includes ND, NE, NA, etc.

The node with the higher priority could choose its beacon slot earlier. Obviously, initiator occupies the first CFBS and its information is broadcasted without the hop limitation. Then other nodes, according to their priorities, can take the slots that are not used by their 2-hop neighbours and store the slot number in their NT. As each node only maintains a 2-hop NT, the nodes may choose CFBS synchronously and therefore quickly. With BOPL, each node can calculate the beginning of a superframe by the time of a received neighbour beacon and its slot number. At this moment, the node synchronizes with its neighbours and can schedule its own superframe. We can say that the node is in working stage now.

More details of CFBS mechanism have been published in [24] , including beacon frame format, theoretical analysis and simulation study of protocol cost such as convergence time and the incurred message overhead.

## 3.3. CFDS Mechanism

For the time-bounded traffic or the traffic of zero-tolerance packet loss, CFDS enables these traffics to be sent in some dedicated slots. Medium access can be done directly, without backoff delays or medium sensing. This aspect is very useful in our application.

Thanks to CFBS, the CFDS negotiations can be achieved between source node and destination node using beacon exchanges. Since the beacons are broadcasted with guarantee through the CFBS, the CFDS allocation process can be time-bounded, unlike the GTS_request process of IEEE 802.15.4 that uses CSMA/CA. Similarly, nodes must determine their own CFDSs by taking into account CFDSs used by their 2-hop neighbourhoods.

For the source node, CFDS allocation is triggered by the upper layer, which contains the destination address and the length of requested CFDS. By setting flags of beacons, source node broadcasts this request to all its neighbours. When a source node receives a beacon from a destination node with the valid CFDS slot number and CFDS length, it will update the NT and broadcast the allocated slot in the next beacon. The new time schedule is also calculated in order to achieve this direct collision-free transmission. As there is no extra overhead for the CFDS negotiation, source node can continue requesting until the destination node finds the valid CFDS.

So for destination node, the objective is to search the available CFDS and respond to source node. The slot number must be decided by the receiver (destination) of the traffic to avoid frame collision. Specifically, when a node receives neighbour's beacon and finds its address as the destination, it will check its NT, allocate the first available data slot (e.g. slot 8 in Figure 1) to source node and announce this allocation in the next beacon. This available CFDS should not be reused by another node in the 2-hop neighbourhood. When source node requests several CFDSs, the slot number is noted as 8 if the following CFDSs are also available. Else, destination node searches until the end of active period (slot 15). If the available CFDSs are not enough, destination node also returns to the first available CFDS and provides CFDS services as more as possible.

The CFDS deallocation is also invoked by source node of this CFDS. Source node clears up all the flags and slot number and then sends the beacon. Destination node also clears up all the CFDS information with this source node from its NT when it receives the indicated beacon. Therefore, the deallocation is complete and the CFDS is free for other nodes.

CFDS mechanism allows point-to-point bidirectional communications in the mesh topology. When a destination node wishes to reserve CFDS with the corresponding source node, it launches the same procedure as single-direction communication. However, considering the indoor application environment and the possible large overhead (a 3-hop NT), the traffic sending by CFDS has no acknowledgement. This specific point has been successfully studied on our real testbed. In addition, if no CFDS used by the node at all, it may sleep during these data slots for further energy saving.

## 3.4. Smart Repair Protocol

The objective of this protocol is to minimize the impact of a change of topology as much as possible. The SRP (Smart Repair Protocol) is critical in ADCF as it improves network flexibility and robustness. Generally, topology changes could be classified as two types: node join and node failure. All other complex topological changes such as network separation and network integration are all based on the two basic changes.

In reality, node failure can be caused by energy exhaustion, unexpected damage or even the poor quality of wireless link. So a node considers the failure of its neighbour node only when the neighbour's beacon loss is above a threshold. Similarly, new node is regarded as neighbour only when its beacon loss is below a threshold. In our proposal, beacon loss threshold is a predefined parameter which depends on different wireless environments during the implementation. In other words, we estimate the wireless link quality by beacon loss. Four states are defined for a link or a node connected by the link. 1) Preliminary. 2) Unconfirmed. 3) Confirmed. 4) Deleted. Therefore, node join or node failure here means the presence or disappearance of one node in its neighbours' NT. Both unconfirmed nodes and confirmed nodes are stored as 1-hop neighbours in NT. But only confirmed nodes are chosen as 1-hop neighbours when constructing beacon frame.

Asymmetric links are possible and acceptable. In ADCF, both CFBS and CFDS mechanisms are over a 2-hop NT. If two neighbours connected with an asymmetric link choose the different slot to transmit the beacon or data frame, the collision could be avoided and that's what we expect. If two neighbours connected with an asymmetric link choose the same slot, they will transmit the beacon or data frame at approximate the same time. In this situation, this asymmetric link will be abandoned and other high quality link may be utilized. The redundant mesh link of ADCF also improves the communication reliability.

Based on the above definitions, the following two parts specify different mechanisms depending on different topology change cases.

1) Node join and BOP augmentation. Except full mesh network, multiple-hop network may reuse the medium so that there are always free slots in BOP. A new node may choose its CFBS directly after the listening period and join the network immediately. So ADCF has the great advantage to adapt the topology change. For full mesh network, if a new node wants to join the network and finds BOPL is not sufficient ($D_{max}$ > BOPL), the new node may send its beacon by CSMA until a new initiator is designed with updated $D_{max}$. In this process, the nodes will return to initialization stage.

2) Node failure and BOP reduction. If link failure is detected by beacon loss mechanism, neighbours will simply delete this failing node from NT. If the initiator fails, others re-select an initiator but keep their BOP with the original slots. Therefore, the network will still work without disruption. In one case ($D_{max}$ > BOPL/2), the nodes will also return to initialization stage for an optimal BOP.

In conclusion, our application requires an adaptive communication protocol providing QoS-guaranteed service with reasonable energy consumption in a mesh network. With ADCF, we proposed two mechanisms called CFBS and CFDS to enable efficient deterministic medium access for beacon and data. The Smart Repair Protocol may further improve the network flexibility and robustness. In the following parts, we will present the performance results obtained by the simulation and the prototyping work.

# 4. SIMULATION STUDY

To study the scope of our contribution, we use OPNET to develop a simulation model that implements the ADCF MAC. Many scenarios and metrics were simulated. For example, we investigated the protocol cost of initialization stage, the network performance with different topology change cases, and the network performance in large scale and high-density situations, etc. [25]. In this paper, we choose to present the comparison of ADCF with IEEE 802.15.4 (ZigBee version implementation [26]). An ideal IEEE 802.15.4 physical layer, indicating the channel without random packet loss, is applied for both protocols. A more practical energy model [27] is used in the simulation. A static routing mechanism achieved by prior manually adding routes is above ADCF in order to simulate application traffics over the network. Buffers sizes are consistent with the prototype. In addition, the topology is a random mesh configuration. Other basic parameters are shown in the Table 1.

According to our energy model, the actual energy consumption in reception mode is larger than the energy consumption in transmission mode. Therefore, even though there is no application traffic, the active period always keeps in receive mode, which consumes more energy. So only beacons are delivered in order to compare the protocols cost. As shown in Figure 2, it can be seen that ADCF consumes less energy, about 37%, compared to IEEE 802.15.4 as time goes by. This is because the active period is the same for all the ADCF nodes, while IEEE 802.15.4 router nodes have two active periods. One is for the communication with parents and the other is for the communication with children. Hence IEEE 802.15.4 spends more time for idle listening. In theory, no matter what topology, if BOPL is less than the active period, ADCF always consumes less energy than the IEEE 802.15.4 standard.

As shown in Figure 3, there are 7 sources with 1-hop traffic or 3 sources with multi-hop traffic. When Packet Interarrival Time decreases from 1.0 s to 0.1 s, the traffic load will increase. Therefore, End-to-end delay becomes larger. When Packet Interarrival Time is about 0.4 s, there are radical changes caused by buffer overflow.

With the current superframe (2 s) and active period configurations, the end-to-end delay for 1-hop traffic is about 1 s on average, including the time from packet generation to the scheduled data slot. ADCF has 0.06 s advantage than IEEE 802.15.4. If a packet arrives just before its scheduled data slot, the node can send this packet immediately. Otherwise, the node may send the packet in the next superframe. It also can be seen that ADCF saves about 25% end-to-end delay for multi-hop traffic. This is because some multi-hop traffic may be transmitted in one superframe as the same active period for all the mesh nodes. IEEE 802.15.4 works in a cluster-tree topology which may take several superframes from the source to the final destination. The average hop count is 3, so this end-to-end delay is about 2.7 s for ADCF and 3.5 s for IEEE 802.15.4. In some other topology cases, ADCF can be even more efficient since the 802.15.4 tree can be shortened thanks to the mesh links.

In the same simulation, when the CFDS buffers are available, both ADCF and IEEE 802.15.4 keep 100% packet success ratio, which indicates the percentage of successfully delivered packets.

**Table 1**. Basic simulation parameters.

| Parameter | Value |
|---|---|
| Scene area | $100 \times 100\ m^{-2}$ |
| Transmission range | 15 m |
| Number of nodes | 14 |
| Superframe period | 2 s |
| Active period | 0.245 s |
| Application payload | 100 bits |
| CSMA buffer | 0.5 k octets |
| CFDS buffer | 1.5 k octets |
| Simulation duration | 30 min |
| Simulation times | 20 |

By many other simulation works and results, we can conclude that ADCF satisfies our application request of delivering QoS messages with low energy consumption. Therefore, for the monitored people, environmental alerts, and remote assistant messages could be quickly sent to the destination without loss. The end-to-end delay is optimized, especially for the multi-hop alters traffic.

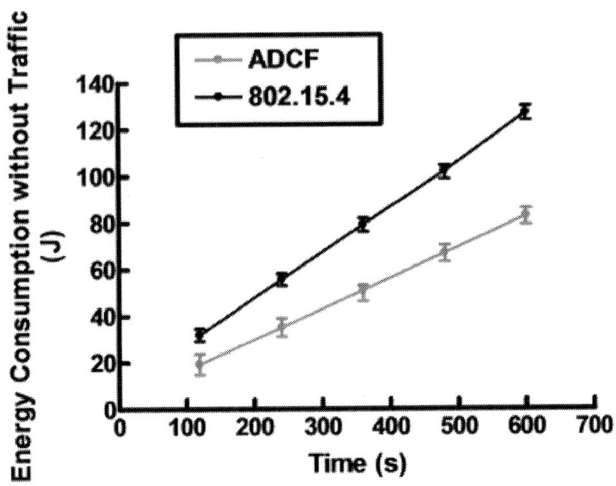


**Figure 2**. Energy consumption comparison.

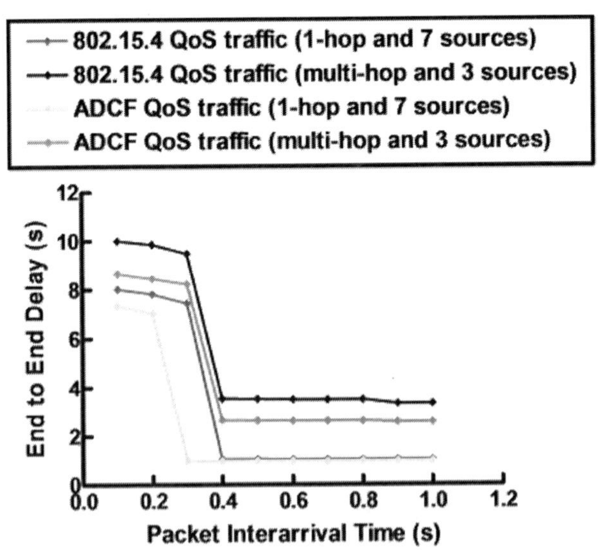


**Figure 3**. End-to-end delay comparison of traffic sent by CFDS.

# 5. PROTOTYPE IMPLEMENTATION

Prototype implementation is a fundamental approach to verify a protocol and its performances. A platform WiNo [28] of emulation and rapid prototyping for WSN is utilized. WiNo is an open platform ready to accommodate protocols at MAC layer or network layer for sensor networks. It provides an open environment adapted to research projects, including the management of physical layer and the necessary tools to develop a full but very compact protocol stack using ANSI C-language. More accurately, a developer with WiNo can master not only the time access to the medium and the sleep-wake cycle, but also the CPU time and memory resource generally restricted by hardware in a WSN.

Currently, two types of nodes from Freescale, 13192-SARD and 1321x-SRB, are implemented with ADCF MAC. Both of them have the common features: 1) IEEE 802.15.4 2.4 GHz transceiver. 2) 8-bit MCU. 3) There are 4KB of RAM and 60KB of on-chip Flash. Two tools are mainly used for debugging, measuring and evaluating the protocol performance in real conditions: the date-logger console of nodes and SNA (Sensor Network Analyzer) [29] . Daintree's SNA is known as an expert tool providing comprehensive solution for IEEE 802.15.4 and ZigBee. It includes a protocol analyzer software fitted to sensor networks and a sensor network adapter hardware. The sensor network adapter hardware is a powerful node, which can capture all the packets of a network but do not participate in this network. By SNA software that includes a protocol decoder allowing user to drill down to packet, field and byte level, we can view all network nodes and interactions simultaneously, without disturbing the original network.

In part 5.1, two results obtained through practical measurements in our laboratory will be given. In part 5.2, the deployment of ADCF in a real application context using our smart home of Blagnac [30] will be presented.

## 5.1. Experimental Scenarios and Results

To build the multi-hop network, 8 nodes are gradually deployed in our laboratory. In each scenario we start each node at different time with a random MAC address. In fact, each multi-hop network is unique and has its own particularities. We do 20 times manipulations separately for the following two networks examples, as shown in Figure 4 and Figure 5. In the first deployment, 8 nodes are disposed throughout the first floor of our laboratory. And in the second deployment, 4 nodes are disposed in the first floor, 3 nodes are disposed in the ground floor, 1 node is disposed on stairway. Even though the fixed positions, network topology may change due to the wireless links. Based on the SNA results, the maximum hop count is 3 in the two network examples. Some interesting results are shown in Table 2.

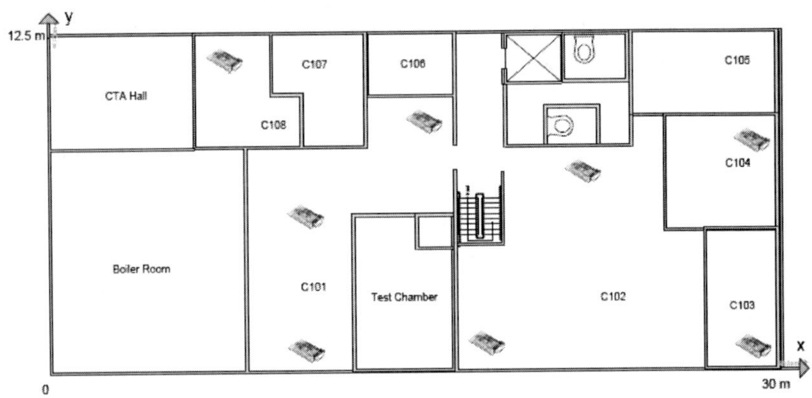


**Figure 4**. Network example 1.

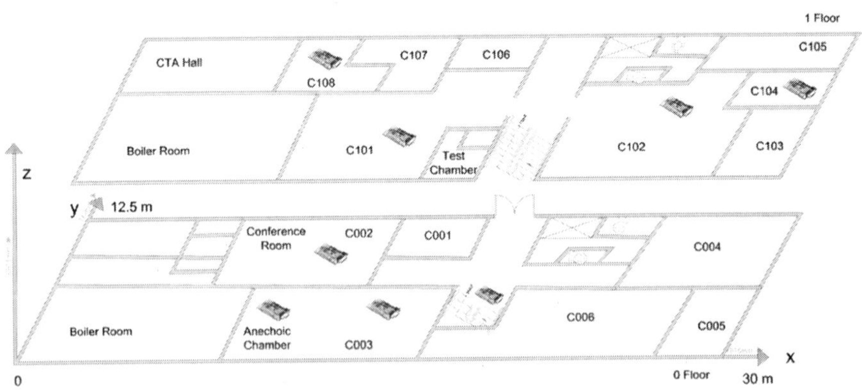


**Figure 5**. Network example 2.

Here convergence time of the last node means the time duration from the start of the last node to a BOP well organized network in which each node has a collision-free beacon slot. In reality for the deployment of multi-hop topology, power-up of each node in the network is a process. At the same time, as in the mesh network each node can directly communicate with its neighbours, therefore network rebuilding may be triggered in this deployment process. We can see that this convergence time takes about 1 to 2 minutes on average. This value is acceptable considering the total network lifetime which is several days or weeks. In a normal situation where the nodes are fixed, the network topology may not be reorganized frequently. The reuse of CFBS confirms the successful achievement of the organized superframe and the synchronized multi-hop networks. In addition, this test highlights that asymmetric link occurs frequently in the real environment, but ADCF can work well without disturbing of asymmetric link problem.

## 5.2. Deployment of ADCF in a Smart Home

The smart home of Blagnac in France targets the elderly and the disabled living alone and provides them health and medico-social assistance at home [31] . Many types of equipment, including wired products, mainly KNX- based, have been installed and are operational in this smart home. ADCF network fills some application gaps and provides an alternative wireless solution. Our testbed is also described in a video [32].

**Table 2**. Results for multi-hop network.

| | | Network example 1 | Network example 2 |
|---|---|---|---|
| Convergence time of the last node | Average | 123.99 s | 97.74 s |
| | Minimum | 55.04 s | 42.63 s |
| | Maximum | 252.30 s | 173.25 s |
| Reuse 1 CFBS | | 19/20 | 11/20 |
| Reuse 2 CFBS | | 1/20 | 9/20 |
| Asymmetric links | | 3/20 | 18/20 |

Six ADCF nodes were deployed in the smart home. From MAC point of view, these 6 nodes form a full mesh network. The nodes are connected with various sensors corresponding to the target of our application:

- Node @106 connected with a magnetic sensor can monitor the open/closure of the refrigerator,
- Node @107 connected with an infrared sensor can detect the motion of the person under his coverage area,
- Node @108 connected with an emergency button is worn by the user and may alert in case of fall, faintness, etc.
- Node @109 connected with a sensor carpet placed near the bed can detect the getting up of the user,
- Node @ 10A connected with a light can be switch on/off by an order of another ADCF node,
- Finally node @105 connected with screen is a sink for collecting and displaying all the network information. On the screen, we can see the superframe structure updating each second and log files from a web page [33].
- From application point of view, the data exchanges are as following (Figure 6):
- Node @106 sends a message to node @105 when the fridge is opened or closed,
- Node @107 sends a message to node @105 when the person moves under the infrared sensor,
- Node @108 sends a message to node @105 when the person presses the emergency button,
- Node @109 sends a message to node @ 10A when the user arrives or leaves the carpet,
- Node @ 10A receives messages from node @109 and switches on the lighting when the user is on the carpet. It switches off the lighting when the user leaves the carpet. When the lighting is switched on/off, node @ 10A sends a message to node @105 containing a return state of lighting.

- Node @105 receives messages from node @106, @107, @108, @ 10A and logs actions in a journal. A voice synthesis software eSpeak [34] is also available and announces the reception of information from sensors (motion, carpet, etc.). Several scenarios using the sensor information have been implemented; for example, when the sink detects the opening of the fridge for more than 10 seconds, the user can hear a voice alert.

In the initialization stage or rebuilding stage, frames are transmitted immediately without any medium access control precaution. In working stage, frames are transmitted using CFDS.

Figure 7 shows a result of the above network deployment. From the ADCF journal and application journal displayed on the screen, we can see that each node can join or leave the network freely and the rest of the network works properly. Generally, the network rebuilding time is less than 10 s. At this moment, 6 nodes occupy 6 CFBS as they are all 1-hop neighbours. 5 CFDS are negotiated to transmit the corresponding application data. Thanks to a buzzer available on the nodes, which is activated for each data frame reception, we can verify that the bounded time is verified as expected. For example, end-to-end delay is always less than 2 s when the superframe is fixed as 1.5 s. It satisfies our application requirements. In addition, the superframe with its slot allocation is totally consistent with what we proposed in Section 3.

# 6. CONCLUSIONS AND PERSPECTIVES

This paper presented an original MAC protocol ADCF based on the IEEE 802.15.4 standard. This protocol was designed to build and to maintain a wireless mesh sensor network providing QoS-guaranteed medium access and energy saving solution for home monitoring application. The work was organized in three phases: protocol presentation, implementation and results from simulation and from prototype. In conclusion, the contributions of this paper are:

1) The 2-hop CFBS and CFDS mechanisms were proposed and described. CFBS enables beacon collision to be avoided and therefore efficient multi-hop mesh network to be formed. Thanks to CFBS, CFDS enables the application data to be transmitted in a bounded time without loss. A smart repair protocol was also proposed to improve the

flexibility and robustness of a mesh network. It's efficient to manage BOPL according to the wireless topology changes.

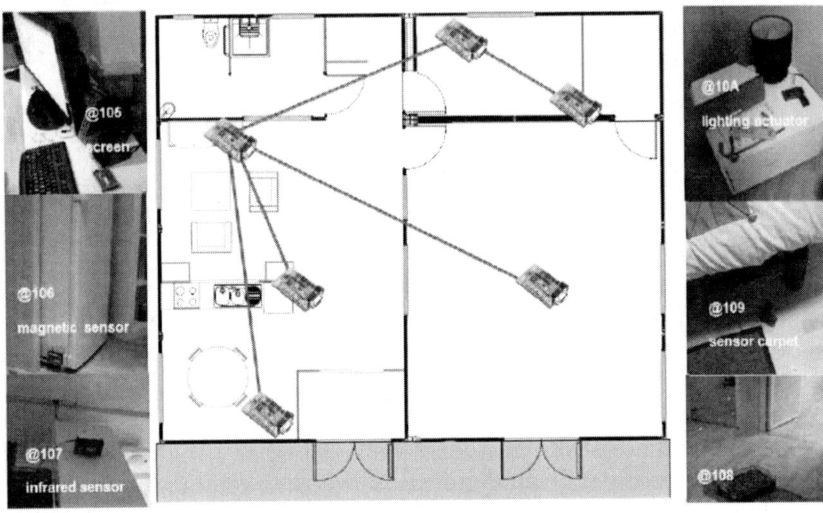


**Figure 6**. Six ADCF sensor nodes and the data transmission topology at application layer.

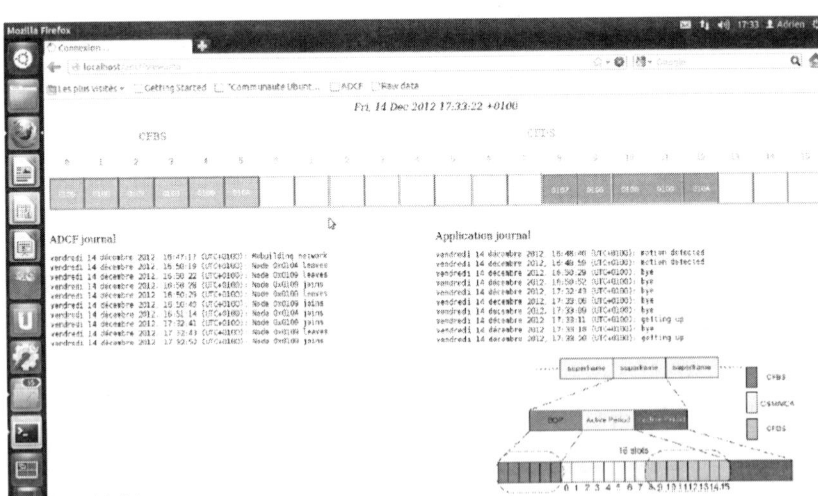


**Figure 7**. Example of a visual result.

2) We established a simulation model by OPNET. The simulation results show that ADCF satisfies our application requirements. For example of the specific topology, ADCF consumes less than 37% energy than IEEE 802.15.4. At the same time, end-to-end delay and packet success ratio show similar performance compared to the standard. Many other scenarios were simulated and the results may be good reference to prototype.

3) We developed some prototypes which were not considered in other works. By the results obtained through practical measurements, we can see: a mesh network of 8 nodes could successfully be built with an organized superframe where each node can choose a CFBS by ADCF. A full mesh network of 6 nodes was deployed in the smart home. This network can always converge in about 10 s with the suitable parameters configuration and, thanks to the buzzers we can hear the reception of an application data on the sink in 2 s at most, which confirms that ADCF is verified in a real environment as expected.

Juan Lu, Adrien Van Den Bossche, Eric Campo, In the future, we will optimize the prototype such as enabling the evaluation of energy consumption of hardware and compare the results with the simulation. Another perspective is in the framework of cross-layer design, such as reuse the 2-hop NT, to reduce routing overhead and further optimize network performances. Finally, we will test various application scenarios and invite the elderly and the disabled to use the system in the smart home.

# REFERENCES


1. Fnet, T. (2006) Counting on an Aging Population. Computing in Science & Engineering, 8, 88-96.

2. Boulanger, J.R. and Deroussent, C. (2008) Preliminary Based Service Evaluation for Elderly People and Healthcare Professional in Residential Home Care Units. Second International Conference on the Digital Society (ICDS 08), 93- 101.

3. Chen, S.-L., Lee, H.-Y., Chen, C.-A., et al. (2009) Wireless Body Sensor Network with Adaptive Low-Power Design for Biometrics and Healthcare Application. IEEE Systems Journal, 3, 398-409.

4. Otal, B., Alonso, L. and Verikoukis, C. (2009) Highly Reliable Energy-Saving Mac for Wireless Body Sensor Networks in Healthcare Systems. IEEE Journal on Selected Areas in Communications, 27, 553-565.

5. Mamaghanian, H., Khaled, N., Atienza, D. and Vandergheynst, P. (2011) Compressed Sensing for Real-Time Energy- Efficient ECG Compression on Wireless Body Sensor Nodes. IEEE Transactions on Biomedical Engineering, 58, 2456-2466.

6. Noury, N., Herve, T., Rialle, V., et al. (2000) Monitoring Behavior in Home Using a Smart Fall Sensor and Position Sensors. 1st Annual International Conference on Microtechnologies in Medicine and Biology (ICMMB 00), 607-610.

7. Zatout, Y. and Campo, E. (2012) Using Wireless Technologies for Healthcare Monitoring at Home: A Survey. IEEE 14th International Conference on e-Health Networking, Applications and Services (HealthCom 12), 383-386.

8. Pomponio, L., Le Goc, M., Pascual, E. and Anfosso, A. (2011) Resident's Activity at Different Abstraction Levels: Proposition of a General Theoretical Framework. IEEE 6th International Conference on Intelligent Data Acquisition and Advanced Computing Systems (IDAACS 11), 540-545.

9. van den Bossche, A., Val, T. and Campo, E. (2007) Prototyping and Performance Analysis of a QoS MAC Layer for Industrial Wireless Network. 7th International Conference on Fieldbuses and Networks in Industrial and Embedded Systems (IFAC 07), 7.

10. Mahfoudh, S. and Minet, P. (2009) Maximization of Energy Efficiency in Wireless Ad Hoc and Sensor Networks with SERENA. Mobile Information Systems, Advances in Wireless Networks, 5, 33-52.

11. IEEE 802.15.4 Standard (2006) Part 15.4: Wireless Medium Access Control (MAC) and Physical Layer (PHY) Speci- fications for Low-Rate Wireless Personal Area Networks (LR-WPANs). IEEE Standard for Information Technology, IEEE-SA Standards Board.

12. Yoo, D.S., Park, S.S., Choi, S.S. and Park, S.H. (2008) Dynamic S-MAC Protocol for Wireless Sensor Networks Based on Network Traffic States. 14th Asia-Pacific Conference on Communications (APCC 2008), Tokyo, 14-16 October 2008, 1-5.

13. Jeong Gil, K., Tsiftes, N., Dunkels, A. and Terzis, A. (2012) Pragmatic Low-Power Interoperability: ContikiMAC vs TinyOS LPL. 9th Annual IEEE Communications Society Conference on Sensor, Mesh and Ad Hoc Communications and Networks (SECON 2012), Seoul, 18-21 June 2012, 94-96.

14. Mihai, G., Alina, D.A. and Ion, B. (2010) Performance Analysis on T-MAC Protocol over a Body Area Network. 3rd International Symposium on Electrical and Electronics Engineering (ISEEE 2010), Galati, 16-18 September 2010, 224-227.

15. Yang, O. and Heinzelman, W.B. (2012) Modeling and Performance Analysis for Duty-Cycled MAC Protocols with Applications to S-MAC and X-MAC. IEEE Transactions on Mobile Computing, 11, 905-921.

16. ZigBee-Alliance. http://www.zigbee.org

17. Koubaa, A., Cunha, A. and Alves, M. (2007) A Time Division Beacon Scheduling Mechanism for IEEE 802.15.4/ Zigbee Cluster-Tree Wireless Sensor Networks. 19th Euromicro Conference on Real-Time Systems (ECRTS 2007), Pisa, 4-6 July 2007, 125-135.

18. Koubâa, A., Alves, M., Attia, M. and Van Nieuwenhuyse, A. (2007) Collision-Free Beacon Scheduling Mechanisms for IEEE 802.15.4/Zigbee Cluster-Tree Wireless Sensor Networks. 7th International Workshop on Applications and Services in Wireless Networks (ASWN 2007), Santander, 24-25 May 2007.

19. Dang, T., Devic, C., Livolant, E., Van Den Bossche, A. and Val, T. (2008) OCARI: Optimization of Communication for Ad Hoc Reliable Industrial Networks. 6th IEEE International Conference on Industrial Informatics (INDIN 2008), Daejeon, 13-16 July 2008, 688-693.

20. Alagha, K., Chalhoub, G., Guitton, A., Livolant, E., Mahfoudh, S., Minet, P., Misson, M., Rahme, J., Val, T. and van den Bossche, A. (2009) Cross-Layering in an Industrial Wireless Sensor Network: Case Study of OCARI. Journal of Networks, 4, 411-420.

21. Muthukumaran, P.S., de Paz, R., Špinar, R. and Pesch, D. (2010) MeshMAC: Enabling Mesh Networking over IEEE 802.15.4 through Distributed Beacon Scheduling. AD HOC Networks, 28, 561-575.

22. Carballido Villaverde, B., De Paz Alberola, R., Rea, S. and Pesch, D. (2010) Experimental Evaluation of Beacon Scheduling Mechanisms for Multihop IEEE 802.15.4 Wireless Sensor Networks. 4th Conference on Sensor Technologies and Applications (SENSORCOMM 2010), Venice, 18-25 July 2010, 226-231.

23. de Paz Alberola, R., Villaverde, B.C. and Pesch, D. (2011) Distributed Duty Cycle Management for IEEE 802.15.4 Beacon-Enabled Wireless Mesh Sensor Networks. IEEE 8th International Conference on Mobile Adhoc and Sensor Systems (MASS 2011), Valencia, 17-22 October 2011, 721-726.

24. Lu, J., van den Bossche, A. and Campo, E. (2011) An Adaptive and Distributed Collision-Free MAC Protocol for Wireless Personal Area Networks. 6th International Symposium on Intelligent Systems Techniques for Ad hoc and Wireless Sensor Networks (IST-AWSN 11), Niagara Falls, 19-21 September 2011, 798-803.

25. Lu, J., van den Bossche, A. and Campo, E. (2012) Improving Robustness and Flexibility of MAC Layer for Guaranteed QoS Indoor Monitoring in Wireless Mesh Sensor Networks. 3rd International Conference on Wireless Communications in Unusual and Confined Areas (ICWCUCA 12), Clermont Ferrand, 28-30 August 2012, 1-6.

26. www.open-zb.net/wpan_simulator.php

27. Fourty, N., van den Bossche, A. and Val, T. (2012) An Advanced Study of Energy Consumption in an IEEE 802.15.4 Based Network: Everything but the Truth on 802.15.4 Node Lifetime. Elsevier, Computer Communications, 35, 1759- 1767.

28. Van den Bossche, A. and Val, T. (2013) WiNo: Une plateforme d'émulation et de prototypage rapide pour l'ingénierie des protocoles en réseaux de capteurs sans fil. UbiMob, Nancy.

29. http://www.daintree.net/sna/sna.php

30. http://mi.iut-blagnac.fr

31. Campo, E., Daran, X. and Redon, L. (2011) A Smart Home between Techniques Sciences and Human Sciences. Une maison intelligente au carrefour des sciences technologiques et des sciences humaines. 2nd Conference on Accessibility and Assistive Systems for Persons in Disabled Situation (ASSISTH 11), 33-42.

32. http://irit.fr/~Adrien.Van-Den-Bossche/adcf_application_demonstration_video_v2.flv

33. http://lab.iut-blagnac.fr/adcf/

34. http://espeak.sourceforge.net/

5

# APPLICATIONS OF WIRELESS SENSOR NETWORKS IN MARINE ENVIRONMENT MONITORING: A SURVEY

**Guobao Xu [1], Weiming Shen [2,*] and Xianbin Wang [2]**

[1]Lab of Ocean Remote Sensing & Information Technology, Guangdong Ocean University, Zhanjiang 524088, China

[2]Department of Electrical and Computer Engineering, The University of Western Ontario, London, Ontario N6A 5B9, Canada

## ABSTRACT

With the rapid development of society and the economy, an increasing number of human activities have gradually destroyed the marine environment. Marine environment monitoring is a vital problem and has increasingly attracted a great deal of research and development attention. During the past decade, various marine environment monitoring systems have been developed. The traditional marine environment monitoring system using an oceanographic research vessel is expensive and time-consuming and has a low resolution both in time and space. Wireless Sensor Networks (WSNs) have recently been considered as potentially promising alternatives for monitoring marine environments since they have a number of advantages such as unmanned operation, easy deployment, real-time monitoring, and relatively low cost. This paper provides a comprehensive review of the state-of-the-art technologies in the field of marine environment monitoring using wireless sensor networks. It first describes application areas, a common architecture of WSN-based oceanographic monitoring systems, a general architecture of an oceanographic sensor node, sensing parameters and sensors, and wireless communication technologies. Then, it presents a detailed review of some related projects, systems, techniques, approaches and algorithms. It also discusses challenges and opportunities in the research, development, and deployment of wireless sensor networks for marine environment monitoring.

## KEYWORDS

wireless sensors; wireless sensor networks; marine environment monitoring; buoys; energy harvesting

## 1. INTRODUCTION

A wireless sensor network (WSN) consists of a number of dedicated sensor nodes with sensing and computing capabilities, which can sense and monitor the physical parameters and transmit the collected data to a central location using wireless communication technologies. A WSN has a number of inherent characteristics including uncontrollable environments, topological constraints, and limited node resources for energy and computational power [1]. Generally, a WSN deploys more sensors than the optimal placement in order to improve the system reliability and the fault tolerance [2].

During the last decade, WSNs have been widely utilized in a variety of application fields related to water monitoring [3–5], forest monitoring [6,7], industrial monitoring [8,9], agriculture monitoring [10,11], battlefield surveillance [12,13], intelligent transportation [14,15], smart homes [16,17], animal behavior monitoring [18,19], and disaster prevention [20,21]. This technology can certainly be applied to the monitoring of marine environments.

On the other hand, with the development of society and economy, more and more people have started to pay attention to the marine environment. Marine environment systems are particularly vulnerable to the effects of human activities related to industry, tourism and urban development [22]. Traditionally, oceanographic research vessels were used to monitor marine environments, which is a very expensive and time-consuming process that has a low resolution both in time and space. For marine environment research, a WSN-based approach can dramatically improve the access to real-time data covering long periods and large geographical areas [23]. According to Tateson *et al.* [24], a WSN-based approach is at least one order of magnitude cheaper than a conventional oceanographic research vessel.

In a WSN-based marine environment monitoring system, various kinds of sensors are used to monitor and measure different physical and chemical parameters such as water temperature, pressure, wind direction, wind speed, salinity, turbidity, pH, oxygen density, and chlorophyll levels.

While the development and deployment of an adaptive, scalable and self-healing WSN system need to address a number of critical challenges such as autonomy, scalability, adaptability, self-healing and simplicity [25,26], the design and deployment of a lasting and scalable WSN for marine environment monitoring should take into account the following challenges different from those on land [22]:

1) *Higher water resistance:* Sensor nodes of a marine monitoring system require greater levels of water resistance;

2) *Stronger robustness:* A marine monitoring system needs stronger robustness, since the marine environment with waves, marine currents, tides, typhoons, vessels, *etc.*, is aggressive and complex, and causes movement of nodes;

3) *Higher energy consumption:* Energy consumption is higher due to long communication distances and an environment in constant motion;

4) *More unstable line-of-sight*: The oscillation of the radio antenna can cause a more unstable line-of-sight between transmitters and receivers [27].

5) *Other problems:* There are also some other problems including the difficulty for deployment and maintenance of nodes, the need for buoy and mooring devices, sensor coverage problems [2], and possible acts of vandalism.

There have been a few literature reviews on Wireless Sensor Networks for marine environment monitoring. Albaladejo *et al.* [22] provided a comprehensive review of the research and development of oceanographic monitoring using wireless sensor networks and pointed out the challenges and difficulties of WSNs for oceanographic monitoring. This paper is intended as an update and extension of Albaladejo *et al*'s review [22] based on recent developments in this area during the past five years. The limitations and challenges of wireless sensor networks for environmental research were discussed in [28]. They reviewed several WSN applications such as water ecosystems, forest monitoring, precision agriculture, wildlife observation, disaster prevention and urban monitoring.

This paper provides a comprehensive review of recent developments in the related fields, discusses major technical challenges, and identifies future research directions. The rest of the paper is organized as follows: Section 2 briefly describes fundamentals of WSN-based marine environment monitoring systems. Section 3 reviews some related projects, systems, and technologies. Section 4 highlights various challenges and opportunities including oceanographic sensors protection, advanced buoy design, energy harvesting system design, and WSN-based system stability and reliability. Section 5 provides some concluding remarks.

# 2. OVERVIEW

This section provides an overview on the application of WSNs in marine environment monitoring, including different application areas, a common architecture of WSN-based marine monitoring systems, a general architecture of an oceanographic sensor node, sensing parameters and sensors, and related wireless communication technologies.

## 2.1. Application Areas

WSN-based marine environment monitoring has a broad coverage including a number of application areas: water quality monitoring, ocean sensing and monitoring, coral reef monitoring, and marine fish farm monitoring. Different application areas require different WSN system architectures, communication technologies, and sensing technologies.

A water quality monitoring system is usually developed to monitor water conditions and qualities including temperature, pH, turbidity, conductivity and dissolved oxygen (DO) for ocean bays, lakes, rivers and other water bodies. An ocean sensing and monitoring system is used to monitor ocean water conditions and other environmental parameters. A coral reef monitoring system is normally installed to monitor coral reef habitats using an autonomous, real-time and *in-situ* wireless sensor network. A marine fish farm monitoring system is developed to monitor water conditions and qualities including temperature and pH, and accurately quantify the amount of fecal waste and uneaten feed for a fish farm.

## 2.2. Common WSN Architecture

Figure 1 shows a common wireless sensor network architecture for monitoring marine environments, which consists of sensor nodes, sink nodes, a base station, a server and user terminals. Sensor nodes can sense and monitor the *in-situ*environmental parameters such as water temperature, salinity, turbidity, pH, oxygen density and chlorophyll levels, and transmit the collected data to sink nodes via wireless communication using ZigBee or some other communication protocol. Communication between sensor nodes and a sink node is usually point-to-point. A sink node collects data from a group of sensor nodes, and transmits the collected data to the base station via the GPRS network. The server stores and processes the received data from the base station. The user terminals connect the server over the Internet.

The design and deployment of a lasting and scalable WSN for marine environment monitoring should carefully take into account the following factors: the hostile environment, the network topology, communication protocols, the number of nodes, buoys, mooring systems, oceanographic sensors, energy supply, and so on.

## 2.3. General Sensor Node

Figure 2 shows an architecture of a general sensor node in a marine environment monitoring system. It usually includes a buoy device in order to protect electronic devices of nodes against water. A marine monitoring sensor node normally consists of the following four main modules [29]:

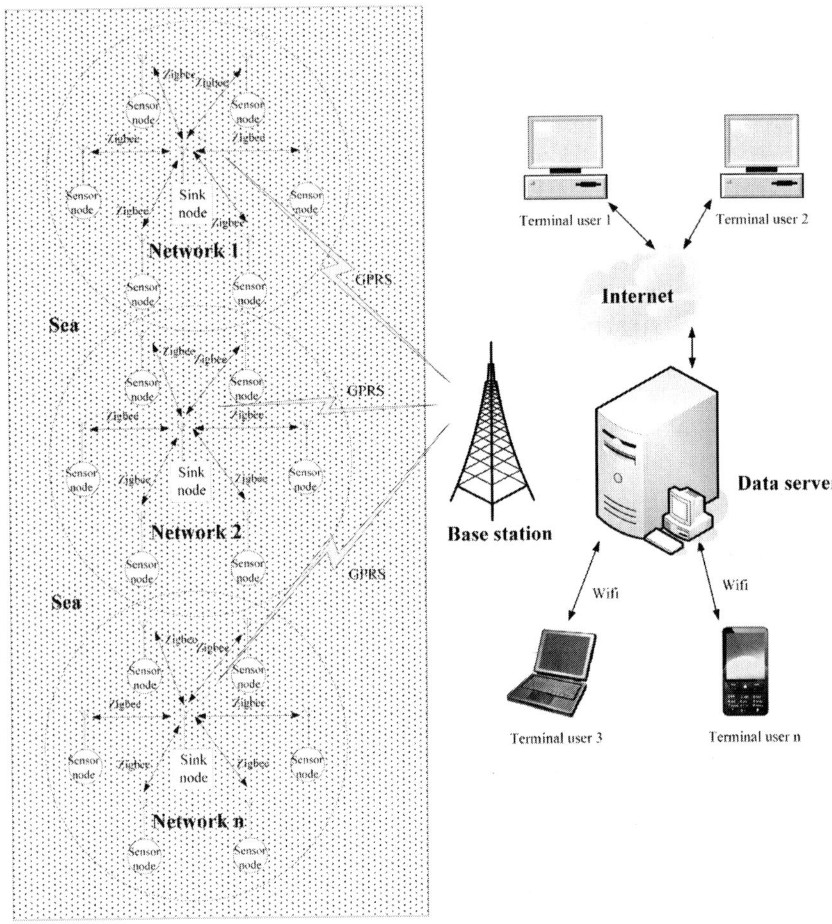


**Figure 1.** Common architecture of WSN-based marine monitoring systems.

1) A sensing module for data acquisition;

2) A central processing module for local data processing and storage;

3) A wireless transceiver module for wireless data communication;

4) A power supply module for energy supply.

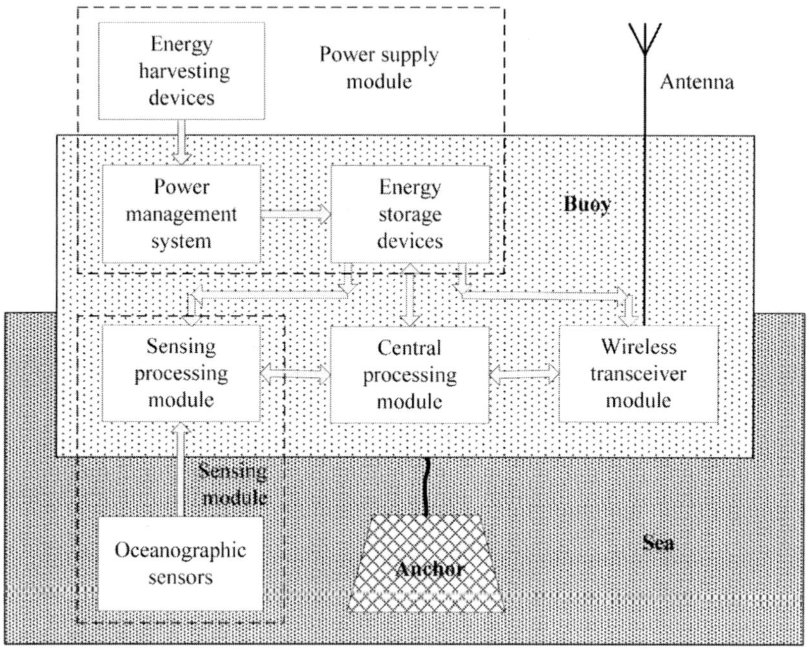


**Figure 2.** General architecture of an oceanographic sensor node.

A sensing module is usually composed of several probes and sensors (with associated amplifiers and A/D converters) to sense and monitor the physicochemical parameters of marine environment as mentioned above. A central processing module normally includes a CPU and memory to process and store the collected data. A wireless transceiver module mainly consists of a RF transceiver and an antenna to send the collected data and receive instructions from the sink node. A power supply module usually contains energy storage devices (rechargeable batteries), power management system and energy harvesting devices (solar panel, wind energy, tidal power, seawater generator, *etc.*). Finally the buoy has an anchor device in order to prevent it from moving (due to waves, marine currents, wind, tide, *etc.*).

The energy options for sensor nodes usually include batteries, capacitors, heat engines, fuel cells, and energy harvesting. Sensor nodes are normally battery powered in most application systems. However, the use of a battery in sensor nodes has a number of disadvantages [30]:

1) As sensor nodes increase in number and size, the replacement of depleted batteries is wasteful and time-consuming.

2) A battery has limited energy that cannot last a long life for sensor nodes.

3) Batteries have environmental contamination and disposal issues since the chemical composition of a battery often involves toxic heavy metals.

It is therefore necessary to explore an alternative power supply for sensor nodes. Harvesting energy from their ambient environment is a promising power supply for sensor networks with lower cost and long life. Energy harvesting methods include photovoltaics, fluid flow, temperature gradients, pressure variations and vibration harvesting. In terms of their efficiencies and realisability, the most outstanding energy harvesting at the moment is photovoltaics [30]. This issue will be further explored in Section 4.3.

## 2.4. Sensing Parameters and Sensors

The operating principle of sensors is to respond to changes in their environment by producing an electrical signal in the form of voltage, current, or frequency [31]. Sensors can commonly be divided into physical sensors and chemical sensors. In a marine monitoring system, physical sensors are used to measure some physical parameters, such as temperature, humidity, pressure, wind speed and wind direction, and chemical sensors are used to sense various chemical parameters (salinity, turbidity, pH, nitrate, chlorophyll, dissolved oxygen (DO), *etc.*) as shown in Table 1.

The right choice of marine environment monitoring sensors depends on the user requirements of deployment area, measurement range, accuracy, resolution, power consumption, and intended deployment time.

**Table 1.** Common marine environment monitoring sensors.

| Sensors | Monitoring Parameters | Range | Accuracy | Power Supply | Unit | Manufacturer |
|---|---|---|---|---|---|---|
| SBE 16plus V2 | Temperature | −5 to +35 °C | ±0.005 °C | 9–28 V | °C | Sea-Bird Electronics |
| GT301 | Pressure | 0 to 60 | < ±0.5% of FRO | 24 V | bar | Kongsberg Maritime |
| SBE 16plus V2 | Conductivity (Salinity) | 0–9 | ± 0.0005 | 9–28 V | S/m | Sea-Bird Electronics |
| OBS-3+ | Turbidity | Mud: 5000–10,000 mg/L<br>Sand: 50,000–100,000 mg/L | 0.5 NTU | 15 V | NTU | Campbell Scientic |
| PS-2102 | pH | 0 to 14 pH | ±0.1 | N/A | pH | PASCO |
| YSI 6025 | Chlorophyll | 0 to 400 μg/L | 0.1 μg/L | 6 V | μg/L | YSI |
| ISUS V3 | Nitrate | 0.007 to 28 mg/L | ±0.028 mg/L | 6–18 V | mg/L | Satlantic |
| SBE 63 | Dissolved oxygen (DO) | 120% of surface saturation in all natural waters | 0.1 | 6–24 V; 35 mA | mg/L | Sea-Bird Electronics |

## 2.5. Wireless Communication Technologies

WSN physical topology and density are entirely dependent on the applications [32], so the design and deployment of a WSN should consider its environment and application. A number of sensor nodes are densely deployed to improve data accuracy and achieve better system connectivity. However, a dense deployment of sensor nodes has some disadvantages: high energy consumption, data collisions, interferences, *etc.* [33]. WSN nodes normally have three typical kinds of network topologies: star topology, cluster/tree topology and mesh topology, as shown in Figure 3.

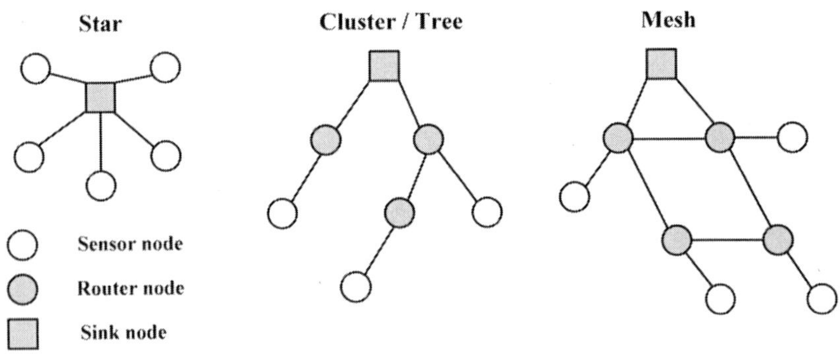


**Figure 3.** General WSN network topologies.

1) *Star topology:* A star topology is a point-to-point single-hop architecture in which each sensor node connects directly to a sink node. It potentially uses the least amount of power among the three topology architectures.

2) *Mesh topology:* A mesh topology is a one-to-many multi-hopping architecture in which each router node connects to multiple nodes. Its advantages over a star topology include a longer range distance of transmission, decreased loss of data, and a higher self-healing communication ability. However, its disadvantages are at the cost of higher latency and higher power consumptions.

3) *Cluster/tree topology:* A cluster/tree topology is a hybrid star–mesh architecture. It takes advantage of the low power consumptions and simple architecture of a star topology, as well as the extended range

and fault tolerance of a mesh one. However, there probably exists some latency.

The right and reasonable choice of network topology depends on the amount and frequency of data to be transmitted, transmission distance, battery life requirements and mobility of the sensor node [34]. It should be noted that a WSN physical topology may change due to available energy, position variations of nodes, malfunction, reachability (due to noise, severe weathers, moving obstacles, *etc.*), and task details of sensor nodes [35].

A sensor node normally incorporates a radio module for wireless communication. The transmitted distance of wireless communication can be anywhere between a few meters (Bluetooth, ZigBee, WiFi, *etc.*) and thousands of kilometers (GSM or GPRS radio communication). Wireless communication has various standards and technologies including Bluetooth, ZigBee, WiFi, GSM, GPRS and WiMAX. Table 2 provides a summary and brief comparison of these communication technologies. Usually, two or more wireless communication technologies are used in a real wireless sensor network. In particular, underwater acoustic communication technologies can be a good choice for data collection and exchange among underwater sensors [36–38].

Generally, the longer the range a radio module must transmit, the more energy consumption a radio module will have. The choice of a wireless communication technology depends on the amount and frequency of the transmitted data, transmission distance, and amount of available energy.

# 3. STATE-OF-THE-ART REVIEW

This section presents a comprehensive review of related projects, systems, applications, network routing mechanisms, algorithms, approaches and techniques on marine environment monitoring based on wireless sensor networks.

**Table 2.** Wireless communication technologies [22].

| Technology | Standard | Description | Throughput | Range | Frequency |
|---|---|---|---|---|---|
| WiFi | IEEE 802.11a; 802.11b/g/n | System of wireless data transmission over computational networks. | 11/54/300 Mbps | <100 m | 5.8 GHz<br>2.4 GHz |
| Bluetooth | IEEE 802.15.1 | Industrial specification for WPAN which enables voice and data transmission between different devices by means of a secure, globally free radio link (2.4 GHz). | v. 1.2: 1 Mbps<br>v. 2.0: 3 Mbps<br>UWB: 53–480 Mbps | Class 1: 100 m<br>Class 2: 15–20 m<br>Class 3: 1 m | 2.4 GHz |
| ZigBee | IEEE 802.15.4 | Specification of a set of high-level wireless communication protocols for use with low consumption digital radios, based on WPAN standard IEEE 802.15.4. | 250 Kbps | <75 m | 2.4 GHz |
| WiMAX | IEEE 802.16 | Standard for data transmission using radio waves. | <75 Mbps | <10 km | 2–11 GHz |
| GSM | | Standard system for communication via mobile telephones incorporating digital technology | 9.6 Kbps | Dependent on service provider | 850/900/1800 /1900 MHz |
| GPRS | | GSM extension for unswitched (or packaged) data transmission. | 56–144 Kbps | Dependent on service provider | 850/900/1800 /1900 MHz |

## 3.1. Related Projects, Systems and Applications

Different WSN projects, systems and applications have been proposed and developed in the literature for monitoring marine environments. Table 3 summarizes the features of related projects, systems and applications.

**Table 3.** Summary of WSN-based marine environment monitoring projects, systems and applications.

| Reference | Organization | Country | Year | Application Areas | Sensing Parameters | Communica-tion Protocols | Buoy | Energy Harvesting | Testing & Deployment | Main Features |
|---|---|---|---|---|---|---|---|---|---|---|
| Perez *et al.* [4] | Universidad Politécnica de Cartagena | Spain | 2011 | Ocean sensing & monitoring | Temperature, pressure, salinity, nitrates, velocity, chlorophyll, and turbidity | GPRS ZigBee | Special buoy | Two solar panels | Deployed in the harbor of Cartagena | LabVIEW-based user interface using Google Maps; Solar energy harvesting; Special buoy |
| Thiemo *et al.* [23] | Swedish Institute of Computer Sci. & University at Berlin | Sweden; Germany | 2007 | Ocean sensing & monitoring | Temperature, motion, vibration and sound | GPRS | Simple buoy and king`s buoy | Batteries | Tested in Baltic Sea | Design of an advanced low-cost buoy system |
| Yang *et al.* [39] | Penn State University | USA | 2002 | Water quality monitoring | pH | RF transceiver and acoustic transducer | PVC housing | Two rechargeabl e batteries | Lab testing with 5 nodes | The design of various interface circuits and the use of five air-based sensor nodes |
| Vesecky *et al.* [40] | UC Santa Cruz | USA | 2007 | Ocean sensing & monitoring | Temperature, wave and location | 900 MHz | mobile minibuoy | Battery power | Prototype buoy tested in a pool | An autonomous mini-buoy prototype; GPS and a PID scheme control |
| Bromage *et al.* [41] | UC Santa Cruz | USA | 2007 | Coral reefs monitoring | Temperature, pH, light, pressure, and conductivity | 900 MHz | Watertight housing | Battery | Monterey Bay deployment | Programmable Oceanic Device (POD) with a 4-mode scheduler to save energy |
| Seders *et al.* [42] | University of Notre Dame | USA | 2007 | Water quality monitoring | Temperature, pH, and DO | 433 MHz | Box and polyethyle ne ring | 12 volt marine battery | Tested a prototype in a small lake | A LakeNet sensor pod and an altered sampling strategy |
| Regan *et al.* [43] | Dublin City University | Ireland | 2009 | Water quality monitoring | Temperature, pH, turbidity, DO and conductivity | ZigBee | Inshore sensor buoys | Solar panel and power pack | Deployed in five sites on the River Lee, Ireland | A real-time heterogeneous water quality monitoring; Sensor maintenance |
| Liu *et al.* [44] | Hong Kong University of Sci. and Tech. | China | 2010 | Ocean sensing & monitoring | Sea depth and temperature | ZigBee | Sensor floating | Batteries | Deployed in HKUST campus and Tsingtao | A Perpendicular Intersection (PI) mobile-assisted localization scheme |
| Lloret *et al.* [45] | Universidad Politecnica de Valencia, | Spain | 2011 | Marine fish farms monitoring | The amount of pollution | ? | Buoy | ? | Tested on OPNET Modeler network simulator | A group-based underwater WSN for monitoring fecal waste and uneaten feed |
| Macias *et al.* [46] | Universidad de Las Palmas de Gran Canaria | Spain | 2011 | Ocean sensing & monitoring | Visible-field, sound and temperature | ZigBee and acoustic | ? | ? | Tested on module of NS-3 | Three tier communication architecture; transmitting video streaming data |
| Roadknight *et al.* [47] | University of Kent | UK | 2004 | Ocean sensing & monitoring | Temperature, conductivity, water depth, turbidity | ? | Single buoy | Batteries | Buoy deployed off Scroby sands | A multi-layered scalable and adaptive approach of data management |
| López *et al.* [48] | Universitat de Barcelona | Spain | 2010 | Fish farm monitoring | Temperature and pH | ZigBee | ? | One rechargeab le battery | Tested in a pool | A sub-layer-based power consumption algorithm |
| O'Connor *et al.* [49] | Dublin City University | Ireland | 2012 | Water quality monitoring | Temperature, conductivity and depth | ? | Buoys | ? | Tested in River Lee, Poolbeg Marina and Galway Bay | A multi-modal environment monitoring network based on WSN and visual image |
| Ceila *et al.* [50] | University of Queensland | Australia | 2009 | Ocean sensing & monitoring | Temperature and illuminance | ZigBee | Cylinder waterproof buoys | Two solar panels | Deployed in the Moreton Bay | Two solar cells and the underwater wireless communication |
| Diofantos *et al.* [51] | Cyprus University of Technology | Cyprus | 2009 | Water quality monitoring | Temperature, pressure salinity and turbidity | GPRS | Cylinder waterproof buoy | Battery | Deployed in a municipal beach | Integrating two technologies of satellite remote sensing and WSN |

| | | | | | | | | | | |
|---|---|---|---|---|---|---|---|---|---|---|
| Yang *et al.* [52] | Zhejiang University of Technology | China | 2009 | Monitoring marine shellfish | Water temperature, pH value, salinity, DO and COD | GPRS | ? | Solar battery | Tested in an aquatic experimental base | Multi-hop communication protocol, multiple nodes, and SMT |
| Jiang *et al* [53] | Ocean University of China | China | 2009 | Ocean Sensing & monitoring | Temperature, velocity and light | ZigBee | Lever buoy | Battery | Deployed off the seashore | The sleep mechanism and lever buoy |
| Jin *et al.* [54] | China Jiliang University | China | 2010 | Water quality monitoring | Temperature, pH, DO, and salinity | ZigBee GPRS | ? | Battery | ? | Two wireless communications of ZigBee and GPRS |
| Chi *et al.* [55] | Shanghai Ocean University | China | 2010 | Ocean Sensing & monitoring | Water temperature, DO and pH | ZigBee | Buoys with GPS &PEA | ? | Experimented in two testbeds | Position determination and location verification using GPS& PEA; Buoys |
| Cesare *et al.* [56] | Politecnico di Milano, Milano | Italy | 2011 | Ocean Sensing & monitoring | Seawater luminosity, temperature and moisture | ZigBee | Cylinder waterproof buoys | Solar energy harvesting | Deployed in the Moreton Bay | Optimal solar energy harvesting; Power-aware and adaptive TDMA protocol |
| De Marziani *et al.* [57] | National University of Patagonia San Juan Bosco | Argentina | 2011 | Ocean Sensing & monitoring | Temperature, pressure, PAR radiation, pH and salinity | ZigBee | Cylinder waterproof buoys | Solar panels | Tested in San Jorge Gulf | A low cost reconfigurable WSN; Buoys; Solar panels |
| Alkandari *et al.* [58] | Kuwait University | Kuwait | 2012 | Water quality monitoring | Water temperature, DO, and pH | ZigBee 802.11 Ethernet radio | ? | A high-capacity solar panel | Tested in a water pool | Using ZigBee and 802.11 Ethernet radio and a high capacity solar panel |
| Albaladejo *et al.* [59] | Technical University of Cartagena | Spain | 2012 | Ocean Sensing & monitoring | Marine temperature and pressure | ZigBee | Special buoy | Solar panels | Deployed in Mar Menor Lagoon | A new multisensory buoy system and solar panels |

Notes: ?: Related information is not available from the reference; DO: Dissolved Oxygen; COD: Chemical Oxygen Demand.

It can be found that most of the efforts are related to general ocean sensing and monitoring [4,23,40,44,46,47,50,53,55–57] and water quality monitoring [39,42,43,49,51,54,58]. Some specific efforts have been made for fish farm monitoring [45,48], coral reef monitoring [41], and marine shellfish monitoring [52]. Some projects focus specific technologies or devices, e.g., buoys [23,40,53,57,59] which will be further discussed in Section 4.2, and energy saving and harvesting [4,43,48,50,52,56–59], discussed in details in Section 4.3.

Most developed systems were only experimented in lab settings or indoor environments [39,45,46,52,55], some are tested in outdoor pools or small ponds/lakes [40,42,48,58], and a number of them have been tested or deployed in real marine or river environments [4,23,41,43,44,47,49–51,53,57,59].

It is also interesting to note that during the last decade most projects, systems and applications have been developed by research groups in a small number of countries, including USA [39,40,41,42], China [44,52–55], Spain [4,45,46,48,59] and Ireland [43,49].

## 3.2. Specific Networks, Routing Mechanisms and Algorithms

To satisfy the requirements for marine environment monitoring systems, researchers have proposed and developed a number of specific networks, routing mechanisms, protocols, and algorithms for WSN-based marine environment monitoring.

A WSN-based data collection framework was proposed and developed by Saha *et al.* [60] for disaster mitigation and rescue operation. A WSN communication protocol with lower delay and better energy efficiency was proposed for data dissemination from disaster areas. A simulation experiment was conducted to validate the performance of the proposed protocol comparing with the SENDROM system protocol.

A group-based underwater wireless sensor network (UWSN) was proposed by Lloret *et al.* [45] to monitor accurately the amount of fecal waste and uneaten feed deposited on the seabed which can cause the damage of the fauna and flora. The design and development of this underwater WSN took into account several factors: number of sensor nodes, sensor nodes mobility model, distribution of sensor nodes, network topology, and communication technologies.

Roadknight *et al.* [47] proposed a multi-layered scalable and adaptive approach of data management for a wireless sensor network. This algorithm consisted of three decision making components: sliding window averaging, local rules and parameter evolution. A single buoy was deployed off Scroby Sands to verify the characteristics of the proposed approach.

A WSN framework was proposed by Lu *et al.* [61] for environmental monitoring applications. Its highlight is on its network layer design by considering multiple aspects: heterogeneity, service-aware control platform, unified routing and scheduling, network monitoring. A special case study was conducted to demonstrate that the framework can be used to guide how to design a WSN for environment monitoring in the future.

Barbosa *et al.* [62] presented a routing algorithm of WSNs for marine oil slick monitoring. They proposed two methods: single relay decision (SRD) and multiple relay decision (MRD) protocols for message routing. The proposed algorithms have more efficient message distribution than single hop and greedy approaches. However, their approach does not consider node mobility, energy harvesting and network scalability.

An IEEE 802.15.4-based wireless monitoring system was presented by Loìpez *et al.* [48] to collect pH and temperature parameters in a fish farm. The proposed algorithm used a ZigBee-based routing and the application layer to manage information transmission from the source node to the central coordinator. They designed a sub-layer-based power consumption algorithm to prolong the node lifetime.

Xu *et al.* [63] proposed an improved WSN MAC protocol for marine environment monitoring to meet the demand of the energy consumption, real-time transmission, bandwidth and reliability. Simulation results show good energy consumption and network throughput abilities. However, the proposed algorithm was not implemented in the actual sensor node to verify its performance.

A WSN dedicated dynamic clustering algorithm was presented for oil slicks monitoring by Harchi *et al.* [64]. It can be applied to a monitoring system adaptively in terms of number of nodes, clustering dynamics, measurement periods, and metric weights to supervise climate conditions. Various parameters are evaluated regarding their influence on the stability of the network clustering algorithm.

Suakanto *et al.* [65] proposed a cloud computing-based approach for data processing in disaster monitoring. The proposed approach used a FTR-HTTP based delivery method from remote client to server.

Jalali *et al.* [66] proposed a cooperative hybrid ARQ (C-HARQ) mechanism in solar powered wireless sensor networks to improve energy efficiency and reliability of energy harvesting. They conducted C-HARQ experiments using a Matlab/Simulink-based simulator for networked and embedded systems. Their experimental results showed that C-HARQ is superior to C-ARQ in energy consumption of relay nodes.

## 3.3. Specific Techniques and Approaches

To address the special needs and purposes of marine environment monitoring, a number of WSN-based techniques and approaches have been developed and reported in the literature.

O'Connor *et al.* [67] presented a multi-modal event monitoring system based on WSNs and visual images for river and coastal marine detection. The system used a visual sensor to complement the use of a WSN in detecting and tracking features of a river or coastal marine location. A software tool was developed to analyze the relationship between the sensor readings and image features. It uses a support vector machine (SVM) approach for training or classification. A Matlab image processing toolbox was used for processing images and extracting various image features including color features, texture features, and edge features.

Kong *et al.* [68] designed a WSN-based water environment monitoring system which can sense and monitor video information in key areas and various water quality parameters, including water temperature, turbidity, pH, dissolved oxygen and electric conductivity. This monitoring system has a data video base station, data monitoring nodes, and a remote monitoring center. This system used an ARM-DSP based double processor, combined ZigBee and CDMA wireless transmission networks, and used a CPLD sampling controller.

A decentralized *ad-hoc* wireless sensor network was proposed for ocean pollution detection by Khan *et al.* [69]. In order to prolong the network lifetime and to improve its Quality of Service (QoS), they focused on the deployment of sensors, protocol stacks, synchronization and the routing algorithms.

A WSN-based wave monitoring technology was proposed by Marin-Perianu *et al.* [70] to monitor various wave parameters. This system deployed dense wireless sensor nodes which are equipped with low-cost, low-power, MEMS-based inertial sensors of accelerometers and gyroscopes. They conducted experiments using a Ferris wheel contraption and the results showed an accuracy of approximately 10 cm for a wheel diameter of 100 cm.

A robotic wireless sensor network was presented by Bhadauria *et al.* [71] for monitoring common carp in Minnesota lakes. This project built a small, mobile, lightweight robotic raft which is deployed with searching and tracking algorithms. They conducted several field experiments in various lakes, and experimental results demonstrated that the robotic raft has great potential in environmental monitoring. They envisioned some system improvements including energy saving, localization accuracy, autonomous navigation and multi-raft systems.

In order to enhance measurement precision and prolong the lifetime of marine environmental monitoring sensors, Delauney *et al.* [72] analyzed the biofouling effects on marine sensors measurements, proposed some promising techniques for the biofouling protection of *in situ* sensors.

To explore the impact of the deep ocean increase in $CO_2$ levels and resulting pH changes on ocean biogeochemistry and ecology, Herlien *et al.* [73] studied a Free Ocean $CO_2$ Enrichment (FOCE) system. The proposed algorithm can achieve the objectives of instrument-in-the-loop control, software reuse of infrastructure and instrument services, and rapid assembly of a scalable end-to-end sensor network system.

An Android-based WSN application was developed by Tembekar [74]. This WSN app can be installed and operate on any Android-based smartphone, get synchronized with the centralized database server, and monitor the various nodes of the wireless sensor network.

# 4. RESEARCH CHALLENGES AND OPPORTUNITIES

So far, wireless sensor networks have been widely applied to terrestrial areas, and some of these deployments have achieved satisfactory performance. However, the application of WSNs in marine environment monitoring is still in its infancy, and most WSN-based systems are purely experimental [22]. This section discusses a few challenges of wireless sensor networks for marine environment monitoring including oceanographic sensors protection, advanced buoy design, energy harvesting system design, and system stability and reliability.

## 4.1. Oceanographic Sensors Protection

In marine environments, there are over 4000 organisms related to fouling problems [75]. According to their sizes, organisms can be classified into micro-organisms (or so called biofilms, slimes, and micro-fouling) and macro-fouling [72]. Biofouling development on a sensor surface is subject to several chemical, physical and biological factors such as pH, dissolved oxygen, temperature, light, location depth, conductivity, organic material and hydrodynamic conditions. When oceanographic sensors are immersed in seawater, they are susceptible to biofouling problems which often lead to the long-term accuracy issues of marine environmental sensor measurements. Since the marine environment is aggressive and the seawater is corrosive, oceanographic sensors should take appropriate fouling protection measures.

The biofouling protection for oceanographic sensors may be divided into three techniques according to their different actions: wipers mechanisms, copper corrosion mechanisms, and chlorine evolution mechanisms [72].

1) *Wiper mechanisms:* A biofouling protection system based on wipers is a purely mechanical method. It is an effective biofouling protection technique as long as the sensor head has a suitable shape for wiper cleaning and the wipers are in good condition.

2) *Copper corrosion mechanisms:* A copper corrosion mechanism is an effective biofouling protection method to protect the sensitive sensor head, but the protection mechanism is not easy to apply to existing sensors and the cost is relatively high.

3) *Chlorine evolution mechanisms:* A biofouling protection system based on a chlorine evolution mechanism uses bleach or chlorine generation by seawater electrolysis. Moreover, this protection mechanism is easily adapted to existing sensors and the cost is relatively low.

Besides the abovementioned three biofouling protection techniques, there are some other interesting methods which promise effective results coming from research laboratories [76,77]. However, it is very difficult to implement these methods in the real sea environment.

Biofouling protection for oceanographic *in-situ* sensors is a very difficult problem. The ideal biofouling protection for oceanographic sensors should take into account six aspects: low cost, low power consumption, easy to install on existing sensors, no or low impact on measurement precision and the environment, long lifetime and robustness against aggressive conditions. Therefore, researchers and manufacturers should further study and explore the biofouling protection mechanisms for marine environmental sensors.

## 4.2. Advanced Buoy Design

Considering the marine environment is aggressive and complex, it is very crucial to design an advanced flotation device (buoy) for a marine environment monitoring system. A buoy normally consists of a wireless sensor network node (CPU, sensors, radio, and batteries), an energy harvesting module, underwater sensors and a mooring system. For example, Pirisi*et al.* [78] proposed a special energy harvesting buoy which can effectively use sea wave energy conversion as a power source and has potential applications in marine environment monitoring. Albaladejo *et al.* [59] designed a multisensory buoy system which can be effectively used for shallow marine environment monitoring.

The design and deployment of an advanced buoy for marine wireless sensor networks should take into account the following requirements: low cost, waterproof, strong stability, energy harvesting, and mooring system.

1) *Low cost:* A marine environment monitoring system using wireless sensor networks is usually composed of a large number of sensor nodes. Therefore, each buoy device needs to be low cost.

2) *Watertightness:* In order to protect the stability of marine environment monitoring system and prolong its lifetime, its electronic devices must be in a waterproof housing to avoid water damage.

3) *Strong stability:* As the marine environment is aggressive and complex, the monitoring system should have a strong stability against adverse atmospheric conditions.

4) *Energy harvesting:* Since it is not convenient to replace the batteries deployed on the marine surface and the sensor nodes, which are far away from the land and are power-hungry, it is necessary to consider the use of energy harvesting to reduce system maintenance requirements.

5) *Mooring system:* Due to tides, waves, marine currents, wind, *etc.*, an anchor is required on the seabed in order to avoid the movement of the buoy devices.

Besides the above mentioned requirements, the buoy mechanic design should meet a number of requirements including the buoy visibility with bright yellow color and a warning light for maritime traffic, the use of environmentally friend materials, the connection of several sensors, and the reasonable antenna height for the better communication propagation.

## 4.3. Energy Harvesting System Design

The energy supply of a wireless sensor network is generally provided by batteries which have limited energy [27]. In addition, in marine environment monitoring systems, wireless sensor nodes are often deployed in unapproachable sea surface areas, and they are mostly planned for long-time operation, therefore, it is not convenient to replace the sensor batteries. Moreover, marine sensor nodes (sink nodes) have high energy consumption due to the use of long-range wireless communication protocol (GPRS). In order to reduce system maintenance requirements effectively, there is a clear need to design an energy harvesting system which uses renewable energies source such as solar [4], tidal power [78], or wind energy [79,80].

Some energy harvesting devices have been designed and developed to prolong the lifetime of marine environment monitoring systems. For example, Perez *et al.* [4] developed a solar energy harvesting device which is composed of two solar panels behind the electronic equipment with an inclination of 45 degrees in an opposite direction. In order that one battery is always being charged while the other is always being discharged, a power management system of a low-power maximum power point tracker (MPPT) circuit was developed and used for wireless sensor networks in [56] and [81].

To design an advanced energy harvesting system for marine environment monitoring, we should consider the following three aspects: energy harvesting devices, power management system, and energy storage devices.

- *Energy harvesting devices:* An energy harvesting device is responsible for harvesting energy from the ambient environment. According to the characteristics of available ambient energies, we should choose appropriate energy harvesting devices and should consider how to install the energy harvesting devices.

- *Power management system:* A power management system can intelligently manage the batteries to be charged and discharged at separate intervals of time. An ideal power management system can prolong the lifetime of batteries and easily store more energy for the system.

- *Energy storage devices:* Energy storage devices normally adopt the rechargeable batteries. Usually, the energy capacity of rechargeable batteries is larger than daily system energy consumption and daily harvesting energy in order to store energy and permit the system to supply power even in case of bad weather [56].

Given the aggressive and hostile marine environment, in order to harvest and use more reliable renewable energies, we can envision a hybrid harvesting energy system for marine environment monitoring in the future, which can use several renewable power sources such as solar, tidal power, seawater generator, and wind energy.

## 4.4. System Stability and Reliability

During the past decade, the system stability and reliability problem of wireless sensor networks has been widely studied in order to measure physical parameters correctly and effectively, as well as to prolong the lifetime of the system dramatically [82,83]. AboElFotoh *et al.* [84] studied the reliability and message delay for cooperative wireless distributed sensor networks subject to random failures. Egeland *et al.* [85] analyzed the reliability and availability of wireless multi-hop networks with stochastic link failures. Qureshi *et al.* [86] presented a methodology using a topology control mechanism for the reliability evaluation of a WSN. Silva *et al.* [8] proposed a methodology based on an automatic generation of a fault tree to evaluate the reliability and availability of wireless sensor networks in typical industrial environments.

Considering the aggressive and complex environment, it is very important to analyze the system reliability in a marine environmental monitoring system using wireless sensor networks. Therefore, the research on the reliability of a WSN-based marine environment monitoring system should take into account the following aspects.

1) *Battery life issues:* As mentioned above, marine sensor nodes (sink nodes) consume more energy than other kinds of wireless sensor nodes. Therefore, the battery life issue always affects the system reliability.

2) *Communication relay issue:* The communication relay affects dramatically the system reliability, when some nodes fail or simply disappear.

3) *Severe environment conditions:* The marine environment always has external interference from ships, fishes, and birds, and has severe weather conditions such as waves, marine currents, tides and typhoons. Such severe environment conditions further influence the system reliability.

## 5. CONCLUSIONS

During the last decade, monitoring of the marine environment has attracted a great deal of research and development attention. Wireless sensor networks are a highly promising technique for monitoring marine environments because of their advantages of easy deployment, real-time monitoring, automatic operation, and low cost. This paper presents a state-of-the-art survey of applications of wireless sensor networks in marine environment monitoring. It first describes fundamentals of WSNs-based marine environment monitoring, including application areas, a common WSN architecture, a general sensor node architecture, sensing parameters and sensors, and wireless communication technologies. Then, it reviews the related literature according to different projects, systems, applications, network routing mechanisms, algorithms, approaches and techniques on marine environment monitoring based on wireless sensor networks. From this survey, it is evident that there are still a few interesting challenges and opportunities on development and deployment of wireless sensor networks for marine environment monitoring, including oceanographic sensors protection, advanced buoy design, energy harvesting system design, and system stability and reliability.

## ACKNOWLEDGMENTS

This work was financially supported by the Guangdong Ocean University scholarship council, and the Guangdong Ocean University Projects (E09174) to establish Ph.D. graduate programs.

# REFERENCES


1. Li, N.; Zhang, N.; Das, S.K.; Thuraisingham, B. Privacy preservation in wireless sensor networks: A state-of-the-art survey. *Ad Hoc Netw.* **2009**, *7*, 1501–1514.

2. Cardei, M.; Wu, J. Energy-efficient coverage problems in wireless ad-hoc sensor networks. *Comput. Commun.* **2006**, *29*, 413–420.

3. Lee, H.C.; Banerjee, A.; Fang, Y.M.; Lee, B.J.; King, C.T. Design of a multifunctional wireless sensor for *in situ*monitoring of debris flows. *IEEE Trans. Instrum. Meas.* **2010**, *59*, 2958–2967.

4. Perez, C.A.; Jimenez, M.; Soto, F.; Torres, R.; López, J.A.; Iborra, A. A system for monitoring marine environments based on Wireless Sensor Networks. Proceedings of the IEEE Conference on OCEANS, Santander, Spain, 6–9 June 2011; pp. 1–6.

5. Jiang, P.; Xia, H.; He, Z.; Wang, Z. Design of a Water Environment Monitoring System Based on Wireless Sensor Networks. *Sensors* **2009**, *9*, 6411–6434.

6. Bayo, A.; Antolín, D.; Medrano, N.; Calvo, B.; Celma, S. Early Detection and Monitoring of Forest Fire with a Wireless Sensor Network System. *Proced. Eng.* **2010**, *5*, 248–251.

7. Yunus, E.A.; Ibrahim, K. Özgür, U. A framework for use of wireless sensor networks in forest fire detection and monitoring. *Comput. Environ. Urban Syst.* **2012**, *36*, 614–625.

8. Silva, I.; Guedes, L.A.; Portugal, P.; Vasques, F. Reliability and Availability Evaluation of Wireless Sensor Networks for Industrial Applications. *Sensors* **2012**, *12*, 806–838.

9. Zhao, G. Wireless Sensor Networks for Industrial Process Monitoring and Control: A Survey. *Netw. Protoc. Algorithms.***2011**, *3*, 46–63.

10. Raul, M.; Samuel, G.M.; Miguel, A.F.; António, L.G.V.; Salviano, F.S.P.S.; Ferreira, P.J.S.G.; Reis, M.J.C.S. Sun, wind and water flow as energy supply for small stationary data acquisition platforms. *Comput. Electron. Agric.* **2008**, *64*, 120–132.

11. Li, X.; Deng, Y.; Ding, L. Study on precision agriculture monitoring framework based on wsn. Proceedings of the 2nd International Conference on Anti-counterfeiting, Security and Identification (ASID 2008), Guiyang, China, 20–23 August 2008; pp. 182–185.

12. Qian, H.; Sun, P.; Rong, Y. Design Proposal of Self-Powered WSN Node for Battle Field Surveillance. *Energy Proced.***2012**, *16*, 753–757.

13. Padmavathi, G.; Shanmugapriya, D.; Kalaivani, M. A Study on Vehicle Detection and Tracking Using Wireless Sensor Networks. *Wirel. Sens. Netw.* **2010**, *2*, 173–185.

14. Tacconi, D.; Miorandi, D.; Carreras, I.; Chiti, F.; Fantacci, R. Using wireless sensor networks to support intelligent transportation systems. *Ad Hoc Netw.* **2010**, *8*, 462–473.

15. Tubaishat, M.; Zhuang, P.; Qi, Q.; Shang, Y. Wireless sensor networks in intelligent transportation systems. *Wirel. Commun. Mob. Comput.* **2009**, *9*, 287–302.

16. Lee, H.; Wu, C.; Aghajan, H. Vision-based user-centric light control for smart environments. *Pervasive Mob. Comput.***2011**, *7*, 223–240.

17. Bangali, J.; Shaligram, A. Energy efficient Smart home based on Wireless Sensor Network using LabVIEW. *Am. J. Eng. Res.* **2013**, *2*, 409–413.

18. Handcock, R.N.; Swain, D.L.; Bishop-Hurley, G.J.; Patison, K.P.; Wark, T.; Valencia, P. Monitoring Animal Behaviour and Environmental Interactions Using Wireless Sensor Networks, GPS Collars and Satellite Remote Sensing. *Sensors***2009**, *9*, 3586–3603.

19. Nadimi, E.S.; Jørgensen, R.N.; Blanes-Vidal, V.; Christensen, S. Monitoring and classifying animal behavior using ZigBee-based mobile ad hoc wireless sensor networks and artificial neural networks. *Comput. Electron. Agric.* **2012**, *82*, 44–54.

20. Bahrepour, M.; Meratnia, N.; Poel, M.; Taghikhaki, Z.; Havinga, P.J.M. Distributed event detection in wireless sensor networks for disaster management. Proceedings of the 2010 2nd International Conference on Intelligent Networking and Collaborative Systems (INCOS), Thessaloniki, Greece, 24–26 November 2010; pp. 507–512.

21. Lacono, M.; Romano, E.; Marrone, S. Adaptive monitoring of marine disasters with intelligent mobile sensor networks. Proceedings of the 2010 IEEE Workshop on Environmental Energy and Structural Monitoring Systems (EESMS), Taranto, Italy, 9 September 2010; pp. 38–45.

22. Albaladejo, C.; Sánchez, P.; Iborra, A.; Soto, F.; López, J.A.; Torres, R. Wireless Sensor Networks for Oceanographic Monitoring: A Systematic Review. *Sensors* **2010**, *10*, 6948–6968.

23. Thiemo, V.; Fredrik, O.S.; Niclas, F. Sensor Networking in Aquatic Environments-Experiences and New Challenges. Proceedings of the 32nd IEEE Conference on Local Computer Networks, Dublin, Ireland, 15–18 October 2007; pp. 793–798.

24. Tateson, J.; Roadknight, C.; Gonzalez, A.; Khan, T.; Fitz, S.; Henning, I. Real World Issues in Deploying a Wireless Sensor Network for Oceanography. Proceedings of the Workshop on Real-World Wireless Sensor Networks, Stockholm, Sweden, 20–21 June 2005; pp. 20–21.

25. Boonma, P.; Suzuki, J. An Adaptive, Scalable and Self-Healing Sensor Network Architecture for Autonomous Coastal Environmental Monitoring. Proceedings of the IEEE Conference on Technologies For Homeland Security, Woburn, MA, USA, 16–17 May 2007; pp. 1–8.

26. Hadim, S.; Mohamed, N. Middleware challenges and approaches for wireless sensor networks. *IEEE Distrib. Syst. Online* **2006**, *7*, 853–865.

27. Alippi, C.; Camplani, R.; Galperti, C.; Roveri, M. Effective design of WSNs: from the lab to the real world. Proceedings of the 3rd International Conference on Sensing Technology, Tainan, Taiwan, 30 November–3 December 2008; pp. 1–9.

28. De la Piedra, A.; Benitez-Capistros, F.; Dominguez, F.; Touhafi, A. Wireless sensor networks for environmental research: A survey on limitations and challenges. Proceedings of the 2013 IEEE Conference on EUROCON, Zagreb, Yugoslavia, 1–4 July 2013; pp. 267–274.

29. Giuseppe, A.; Marco, C.; Mario, D.F.; Andrea, P. Energy conservation in wireless sensor networks: A survey. *Ad Hoc Netw.* **2009**, *7*, 537–568.

30. Knight, C.; Davidson, J.; Behrens, S. Energy Options for Wireless Sensor Nodes. *Sensors* **2008**, *8*, 8037–8066.

31. Porter, J.; Arzberger, P.; Braun, H.W.; Bryant, P.; Gage, S.; Hansen, T.; Hanson, P.; Lin, C.C.; Lin, F.P.; Kratz, T.; Michener, W.; Shapiro, S.; Williams, T. Wireless Sensor Networks for Ecology. *BioScience* **2005**, *55*, 561–572.

32. Li, M.; Yang, B. A Survey on topology issues in wireless sensor networks. Proceedings of the 4th International Symposium on Information Processing in Sensor Networks, Las Vegas, NV, USA, 25–27 April 2005; pp. 1–7.

33. Zeng, Y.; Sreenan, C.J.; Xiong, N.; Yang, L.T.; Park, J.H. Connectivity and coverage maintenance in wireless sensor networks. *J. Supercomput.* **2010**, *52*, 23–46.

34. Flammini, A.; Ferrari, P.; Marioli, D.; Sisinni, E.; Taroni, A. Wired and wireless sensor networks for industrial applications. *Microelectron. J.* **2009**, *40*, 1322–1336.

35. Kim, S.; Guzide, O.; Cook, S. Towards an Optimal Network Topology in Wireless Sensor Networks: A Hybrid Approach. Proceedings of the ISCA First International Conference on Sensor Networks and Applications, San Francisco, CA, USA, 4–6 November 2009; pp. 13–18.

36. Jiang, Z. User Density in a Cluster Underwater Acoustic Network. *Int. J. Intell. Control Syst.* **2012**, *17*, 31–40.

37. Jiang, Z. Underwater Acoustic Networks – Issues and Solutions. *Int. J. Intell. Control Syst.* **2008**, *13*, 152–161.

38. Zhu, Y.; Jiang, Z.; Peng, Z.; Zuba, M.; Cui, J.; Chen, H. Towards Practical MAC Design for Underwater Acoustic Networks. Proceedings of the 32nd IEEE International Conference of Computer Communications (INFOCOM) 2013, Turin, Italy, 14–19 April 2013; pp. 683–691.

39. Yang, X.; Ong, K.G.; Dreschel, W.R.; Zeng, K.; Mungle, C.S.; Grimes, C.A. Design of a wireless sensor network for long-term, in-situ monitoring of an aqueous environment. *Sensors* **2002**, *2*, 455–472.

40. Vesecky, J.F.; Laws, K.; Petersen, S.I.; Bazeghi, C.; Wiberg, D. Prototype autonomous mini-buoy for use in a wireless networked, ocean surface sensor array. Proceedings of the IEEE International Geoscience and Remote Sensing Symposium, Barcelona, Spain, 23–28 July 2007; pp. 4987–4990.

41. Bromage, M.; Obraczka, K.; Potts, D. SEA-LABS: A wireless sensor network for sustained monitoring of coral reefs. Proceedings of the 6th international IFIP-TC6 conference on Ad Hoc and sensor networks, wireless networks, next generation internet, Atlanta, GA, USA, 14–18 May 2007; pp. 1132–1135.

42. Seders, L.A.; Shea, C.A.; Lemmon, M.D.; Maurice, P.A.; Talley, J.W. LakeNet: An Integrated Sensor Network for Environmental Sensing in Lakes. *Environ. Eng. Sci.* **2007**, *24*, 183–191.

43. Regan, F.; Lawlor, A.; Flynn, B.O.; Torres, J.; Martinez-Catala, R.; O'Mathuna, C.; Wallace, J. A demonstration of wireless sensing for long term monitoring of water quality. Proceedings of the IEEE 34th Conference on Local Computer Networks, Zurich, Switzerland, 20–23 October 2009; pp. 819–825.

44. Liu, K.; Yang, Z.; Li, M.; Guo, Z.; Guo, Y.; Hong, F.; Yang, X.; He, Y.; Feng, Y.; Liu, Y. Oceansense: Monitoring the sea with wireless sensor networks. *Mob. Comput. Commun. Rev.* **2010**, *14*, 7–9.

45. Lloret, J.; Sendra, S.; Garcia, M.; Lloret, G. Group-based Underwater Wireless Sensor Network for Marine Fish Farms. Proceedings of the 2011 IEEE GLOBECOM Workshops, Houston, TX, USA, 5–9 December 2011; pp. 115–119.

46. Macias, E.; Suarez, A.; Chiti, F.; Sacco, A.; Fantacci, R. A Hierarchical Communication Architecture for Oceanic Surveillance Applications. *Sensors* **2011**, *11*, 11343–11356.

47. Roadknight, C.; Parrott, L.; Boyd, N.; Marshall, I.W. A Layered Approach to *in situ* Data Management on a Wireless Sensor Network. Proceedings of the International Conference on Intelligent Sensors, Sensor Networks and Information Processing, Melbourne, Australia, 14–17 December 2004; pp. 85–90.

48. Loìpez, M.; Goìmez, J.M.; Sabater, J.; Herms, A. IEEE 802.15.4 based wireless monitoring of pH and temperature in a fish farm. Proceedings of the 15th IEEE Mediterranean Electrotechnical Conference on MELECON 2010, Valletta, Malta, 26–28 April 2010; pp. 575–580.

49. O'Connor, E.; Zhang, D.; Smeaton, A.F.; O'Connor, N.E.; Regan, F. Multi-Modal Sensor Networks for More Effective Sensing in Irish Coastal and Freshwater Environments. Proceedings of the IEEE Oceans, Hampton Roads, VA, USA, 14–19 October 2012; pp. 1–9.

50. Cella, U.M.; Shuley, N.; Johnstone, R. Wireless Sensor Networks in Coastal Marine Environments: a Study Case Outcome. Proceedings of the Fourth ACM International Workshop on UnderWater Networks, Berkeley, CA, USA, 4–6 November 2009; pp. 1–8.

51. Diofantos, G.H.; Marinos, G.H.; Kyriacos, T.; Athos, A. Integration of micro-sensor technology and remote sensing for monitoring coastal water quality in a municipal beach and other areas in Cyprus. Proceedings of the SPIE Remote Sensing for Agriculture, Ecosystems, and Hydrology, Berlin, Germany, 18 September 2009.

52. Yang, H.; Wu, H.; He, Y. Architecture of Wireless Sensor Network for Monitoring Aquatic Environment of Marine Shellfish. Proceedings of the 7th Asian Control Conference, Hong Kong, China, 27–29 August 2009; pp. 1147–1151.

53. Jiang, M.; Guo, Z.; Hong, F.; Ma, Y.; Luo, H. OceanSense: A Practical Wireless Sensor Network on the Surface of the Sea. Proceedings of the IEEE International Conference on Pervasive Computing and Communications, Galveston, TX, USA, 9–13 March 2009; pp. 1–5.

54. Jin, N.; Ma, R.; Lv, Y.; Lou, X.; Wei, Q. A Novel Design of Water Environment Monitoring System Based on WSN. Proceedings of the 2010 International Conference on Computer Design and Applications (ICCDA), Qinhuangdao, China, 25–27 June 2010; pp. 593–597.

55. Chi, T.; Zhang, H.; Chen, M.; Feng, G. Implementation Study of a Home Position Monitoring System for Marine Environment. Proceedings of the 2nd IEEE International Conference on Information Management and Engineering (ICIME), Chengdu, China, 16–18 April 2010; pp. 231–224.

56. Cesare, A.; Romolo, C.; Cristian, G.; Manuel, R. A Robust, Adaptive, Solar-Powered WSN Framework for Aquatic Environmental Monitoring. *IEEE Sens. J.* **2011**, *11*, 45–55.

57. De Marziani, C.; Alcoleas, R.; Colombo, F.; Costa, N.; Pujana, F.; Colombo, A.; Aparicio, J.; Álvarez, F.J.; Jimenez, A.; Ureña, J.; Hernández, A. A low cost reconfigurable sensor network for coastal monitoring. Proceedings of the IEEE Conference on OCEANS, Santander, Spain, 6–9 June 2011; pp. 1–6.

58. Alkandari, A.; Alnasheet, M.; Alabduljader, Y.; Moein, S.M. Water monitoring system using Wireless Sensor Network (WSN): Case study of Kuwait beaches. Proceedings of the Second International Conference on Digital Information Processing and Communications (ICDIPC), Klaipeda City, Lithuania, 10–12 July 2012; pp. 10–15.

59. Albaladejo, C.; Soto, F.; Torres, R.; Sánchez, P.; López, J.A. A low-cost sensor buoy system for monitoring shallow marine environments. *Sensors* **2012**, *12*, 9613–9634.

60. Saha, S.; Matsumoto, M. A Framework for Disaster Management System and WSN Protocol for Rescue Operation. Proceedings of the IEEE Region 10 Conference on TENCON 2007, Taipei, Taiwan, 30 October–2 November 2007; pp. 1–4.

61. Lu, K.; Qian, Y.; Rodriguez, D.; Rivera, W.; Rodriguez, M. Wireless Sensor Networks for Environmental Monitoring Applications: A Design Framework. Proceedings of the Global Communications Conference, Washington, DC, USA, 26–30 November 2007; pp. 1108–1112.

62. Barbosa, P.; White, N.M.; Harris, N.R. Wireless Sensor Network for Localized Maritime Monitoring. Proceedings of the 22nd International Conference on Advanced Information Networking and Applications, Okinawa, Japan, 25–28 March 2008; pp. 681–686.

63. Xu, L.; Zhang, J.; Li, H.H.; Ye, P.; Yan, X.; Li, X. An Improved MAC Protocal for Marine-Environment Monitoring WSN System. *J. Netw.* **2012**, *7*, 1900–1907.

64. Harchi, S.; Georges, J.; Divoux, T. WSN dynamic clustering for oil slicks monitoring. Proceedings of the 2012 International Conference on Wireless Communications in Unusual and Confined Areas (ICWCUCA), Clermont Ferrand, France, 28–30 August 2012; pp. 1–6.

65. Suakanto, S.; Supangkat, S.H.; Suhardi; Saragih, R.; Nugroho, T.A.; Nugraha, I.G.B.B. Environmental and Disaster Sensing Using Cloud Computing Infrastructure. Proceedings of the 2012 International Conference on Cloud Computing and Social Networking, Bandung, West Java, 26–27 April 2012; pp. 1–6.

66. Jalali, F.; Khodadoustan, S.; Ejlali, A. Cooperative Hybrid ARQ in Solar Powered Wireless Sensor Networks.*Microelectron. Reliab.* **2012**, *52*, 3043–3052.

67. O'Connor, E.; Smeaton, A.F.; O'Connor, N.E. A multi-modal event detection system for river and coastal marine monitoring applications. Proceedings of the 2011 IEEE OCEANS, Santander, Spain, 6–9 June 2011; pp. 1–10.

68. Kong, Y.; Jiang, P. Development of Data Video Base Station in Water Environment Monitoring Oriented Wireless Sensor Networks. Proceedings of the International Conference on Embedded Software and Systems Symposia, Sichuan, China, 29–31 July 2008; pp. 281–286.

69. Khan, A.; Jenkins, L. Undersea wireless sensor network for ocean pollution prevention. Proceedings of the 3rd International Conference on Communication Systems Software and Middleware and Workshops, COMSWARE 2008, Bangalore, India, 6–10 January 2008; pp. 2–8.

70. Marin-Perianu, M.; Chatterjea, S.; Marin-Perianu, R.; Bosch, S.; Dulman, S.; Kininmonth, S. Wave Monitoring with Wireless Sensor Networks. Proceedings of the International Conference on Intelligent Sensors, Sensor Networks and Information Processing, Sydney, Australia, 15–18 December 2008; pp. 611–616.

71. Bhadauria, D.; Isler, V.; Studenski, A.; Tokekar, P. A Robotic Sensor Network for Monitoring Carp in Minnesota Lakes. Proceedings of the 2010 IEEE International Conference on Robotics and Automation,, Anchorage, AK, USA, 3–7 May 2010; pp. 3837–3842.

72. Delauney, L.; Compère, C.; Lehaitre, M. Biofouling protection for marine environmental sensors. *Ocean Sci.* **2010**, *6*, 503–511

73. Herlien, R.; O'Reilly, T.; Headley, K.; Edgington, D.R.; Tilak, S.; Fountain, T.; Shin, P. An ocean observatory sensor network

application. Proceedings of the 2010 IEEE Conference on Sensors, Kona, HI, USA, 1–4 November 2010; pp. 1837–1842.

74. Tembekar, S.; Saxena, A. Monitoring Wireless Sensor Network using Android based SmartPhone Application. *IOSR J. Comput. Eng.* **2014**, *16*, 53–57.

75. Yebra, D.M.; Kiil, S.; Dam-Johansen, K. Antifouling technology—past, present and future steps towards efficient and environmentally friendly antifouling coatings. *Prog. Org. Coat.* **2004**, *50*, 75–104.

76. Whelan, A.; Regan, F. Antifouling strategies for marine and riverine sensors. *J. Environ. Monit.* **2006**, *8*, 880–886.

77. Manov, D.V.; Chang, G.C.; Dickey, T.D. Methods for reducing biofouling of moored optical sensors. *J. Atmos. Ocean. Technol.* **2004**, *21*, 958–968.

78. Pirisi, A.; Grimaccia, F.; Mussetta, M.; Zich, R.E.; Johnstone, R.; Palaniswami, M. Optimization of an Energy Harvesting Buoy for Coral Reef Monitoring. Proceedings of the 2013 IEEE Congress on Evolutionary Computation, Cancún, Mexico, 20–23 June 2013; pp. 629–634.

79. Hsu, C.L. Constructing transmitting interface of running parameters of small-scaled wind-power electricity generator with WSN modules. *Expert Syst. Appl.* **2010**, *37*, 3893–3909.

80. Seah, W.K.G.; Zhi, A.E.; Tan, H. Wireless sensor networks powered by ambient energy harvesting (WSN-HEAP)—Survey and challenges. Proceedings of the 1st International Conference on Wireless Communication, Vehicular Technology, Information Theory and Aerospace & Electronic Systems Technology, Aalborg, Denmark, 17–20 May 2009; pp. 1–5.

81. Cesare, A.; Cristian, G. An adaptive system for optimal solar energy harvesting in wireless sensor network nodes.*IEEE Trans. Circuits Syst. I Regul. Pap.* **2008**, *55*, 1742–1750.

82. Islam, K.; Shen, W.; Wang, X. Wireless Sensor Network Reliability and Security in Factory Automation: A Survey.*IEEE Trans. Syst. Man Cybern. Part C.* **2012**.

83. Shen, Z.; Man, K.L.; Lei, C.U.; Lim, E.G.; Choi, J. Assuring System Reliability in Wireless Sensor Networks Via Verification and Validation. Proceedings of the 2012 International on SoC Design Conference (ISOCC), Jeju Island, Korea, 4–7 November 2012; pp. 285–288.

84. AboElFotoh, H.M.F.; Iyengar, S.S.; Chakrabarty, K. Computing reliability and message delay for cooperative wireless distributed sensor networks subject to random failures. *IEEE Trans. Reliab.* **2005**, *54*, 145–155.

85. Egeland, G.; Engelstad, P. The availability and reliability of wireless multi-hop networks with stochastic link failures.*IEEE J. Sel. Areas Commun.* **2009**, *27*, 1132–1146.

86. Qureshi, H.K.; Rizvi, S.; Saleem, M.; Khayam, S.A.; Rakocevic, V.; Rajarajan, M. Poly: A reliable and energy efficient topology control protocol for wireless sensor networks. *Comput. Commun.* **2011**, *34*, 1235–1242.

6

# RELATIVE LOCALIZATION IN WIRELESS SENSOR NETWORKS FOR MEASUREMENT OF ELECTRIC FIELDS UNDER HVDC TRANSMISSION LINES

**Yong Cui [1,*], Qiusheng Wang [1], Haiwen Yuan [1], Xiao Song [1], Xuemin Hu [1] and Luxing Zhao [2]**

[1]School of Automation Science and Electrical Engineering, Beihang University, Beijing 100191, China

[2]High Voltage Department, China Electrical Power Research Institute, Beijing 100192, China

## ABSTRACT

In the wireless sensor networks (WSNs) for electric field measurement system under the High-Voltage Direct Current (HVDC) transmission lines, it is necessary to obtain the electric field distribution with multiple sensors. The location information of each sensor is essential to the correct analysis of measurement results. Compared with the existing approach which gathers the location information by manually labelling sensors during deployment, the automatic localization can reduce the workload and improve the measurement efficiency. A novel and practical range-free localization algorithm for the localization of one-dimensional linear topology wireless networks in the electric field measurement system is presented. The algorithm utilizes unknown nodes' neighbor lists based on the Received Signal Strength Indicator (RSSI) values to determine the relative locations of nodes. The algorithm is able to handle the exceptional situation of the output permutation which can effectively improve the accuracy of localization. The performance of this algorithm under real circumstances has been evaluated through several experiments with different numbers of nodes and different node deployments in the China State Grid HVDC test base. Results show that the proposed algorithm achieves an accuracy of over 96% under different conditions.

## KEYWORDS

wireless sensor networks; electric field measurement; relative localization; Received Signal Strength Indicator-based (RSSI-based)

## 1. INTRODUCTION

High Voltage Direct Current (HVDC) transmission systems are more favorable than the conventional High Voltage Alternating Current (HVAC) systems because of their economical, technical, and environmental advantages for long distance and bulk power transmission. To date, a ±800 kV HVDC transmission system has been in commercial operation and a ±1000 kV transmission system is being studied on a research basis in China [1]. The electrical environment under HVDC transmission lines is characterized by several different electrical parameters. These parameters are the electric field, the ion current density, the space charge density, the radio interference, the audible noise and the magnetic field [2–4]. The electric field under transmission lines is one of the important parameters for evaluation of the electromagnetic environment of concern to people. Electric field measurement systems based on Wireless Sensor Networks (WSNs) have been applied extensively in the electrical power system on account of their convenience. During the process of electric field measurement, it is necessary to obtain the electric field distribution at different locations under the transmission lines, therefore, the location information of each sensor node is crucial to the analysis of the measurement results. The main focus of the research presented in this paper will be how to obtain the node location information.

To measure the electric field strength at different positions under transmission lines, electric field sensors are arrayed along the direction perpendicular to the transmission lines. The wireless node is connected with the each electric field sensor through twisted-pair braid shielded cable. The wireless nodes' primary function is to acquire and process the output analog signals of the sensor and send the digital electric field data to the remote computer. The key requirement to interpret the received electric field data is to determine the locations of the electric field sensors. The construction of the wireless nodes can be reduced to a one-dimensional linear topology where nodes have a spatial ordering. This spatial characteristic indicates the relative locations for nodes and can facilitate the establishment of the corresponding relationship between the wireless node and electric field sensor. The relative location relationship among the wireless nodes can be described as a node location permutation. A relative localization method based on Received Signal Strength Indicator (RSSI) is elaborated to obtain the correct node location permutation to get the ordering of the wireless nodes. Compared with the existing approach which gathers the location information by manually labelling sensors during deployment, automatic localization can benefit the deployment in two aspects. Firstly, in the case of a large number of nodes, the relative localization method would be more efficient and less error prone. Secondly, the relative localization method can easily re-establish the ordering in situations where a failed node is replaced or a new node is added to the current network.

The developed localization techniques can be divided into two categories: range-based and range-free [5]. Range-based localization estimates distance or angle between the unknown nodes and the anchor node, while range-free localization exploits radio connectivity to confirm proximity or exploits the sensing capabilities of each sensor. Range-based localization should estimate distance using different methods such as Time of Arrival (TOA) [6], Time Difference of Arrival (TDOA) [7], Angle of Arrival (AOA) [8], and RSSI-based ones [9–13], and then use the trilateration location algorithm, triangulation location algorithm or maximum likelihood estimation method to calculate the location of the node [14,15]. All four distance measurement methods, except for RSSI-based methods, have superior measuring precision, hence range-based localization can accomplish accurate location.

The RSSI-based localization methods can be divided into two kinds: range-based methods and range-free methods. Range-based methods assume that nodes can estimate the distance, which needs extra hardware and thus is not appropriate for large-scale outdoor sensor networks. Range-free methods localize nodes based on simple sensing features, such as wireless connectivity, anchor proximity, or localization events detection [16,17]. Among these, connectivity-based solutions feature a low cost, but at the expense of localization accuracy. Radio Frequency (RF) signals are vulnerable to environmental impact which will lead to the instability of RSSI. The environment under high voltage transmission lines is extremely complex. Electromagnetic interference, topography, vegetation, and many other factors may all have an impact on the RSSI. This will cause a large error when estimating the distance through the direct use of RSSI. Although RSSI is not considered as the best choice for distance estimation in the situation, it does provide some useful distance related information which can be used. The experimental result in the paper confirms that the radio signal strength weakens approximately monotonically with the physical distance and the RSSI could provide potential information about which neighbor node is closer and which is farther. Thus, this paper focuses on the research of range-free methods to obtain the nodes relative location information and accordingly realize the nodes sorting.

In [10], a range-free localization beyond connectivity is presented to capture the relative distance between 1-hop neighboring nodes from their neighborhood orderings that serve as unique high-dimensional location signatures for nodes in the network. The signatures are subsequently utilized to estimate the distance between nodes. In [11], one-dimensional localization algorithm based on RSSI is proposed. The algorithm utilizes the ratios of signal strengths, instead of signal strength, to determine the locations of the unknown nodes. The algorithm only needs to establish the data set for the ratio of the signal strengths once. In [12,13], the proposed algorithms assume that closer nodes receive higher RSSI. In [12], the presented algorithm can only be used in the case of the known coordinates of the database, and the wireless sensor network topology is a linear structure or a mesh structure. In [13], the algorithm that does not require the anchor node is proposed to accomplish the relative localization.

Based on the aforementioned points for network deployment, a novel and practical RSSI-based localization algorithm for WSN which assumes that closer nodes receive higher RSSI is proposed for the implementation of relative localization. The algorithm is able to correct the exceptional situation of nodes in the output location permutation to improve the accuracy of the localization. As our localization algorithm can correct RSSI-based relative localization, it is named as "Correct RSSI-Based Relative for Localization", or "CRBR-Localization" for short.

This paper is organized as follows: in Section 2, the motivation for the CRBR-Localization algorithm is described with empirical data and comparisons with existing methods. In Section 3, the CRBR-Localization algorithm is proposed and elaborated. In Section 4, based on the experimental results, the effectiveness of the algorithm is evaluated in four different deployments. Finally, a conclusion section ends this paper.

# 2. MOTIVATION FOR THE CORRECT RSSI-BASED RELATIVE-LOCALIZATION (CRBR-LOCALIZATION) ALGORITHM

## 2.1. Empirical Data as Motivation

(1.) Received Signal Strength Indicator (RSSI) and physical distance

Theoretically, RSSI monotonically varies with the physical distance between the sender node and the receiver node. To evaluate the relationship of RSSI and physical distance, outdoor experiments were conducted in three types of environment: a park lawn, a school playground and the China National Grid HVDC test base. In the experiments, 10 sender nodes were arranged at different distances from the receiver node. All the nodes are located in a straight line and the interval of the sender node is 2 m. Each sender node broadcasted 19-bytes test data packets with 18 dB·m sending power, and the receiver node records the RSSI upon receiving the packet. The test is conducted 100 times and the average value of RSSI of each node from different distance is calculated. The experimental results are shown in Figure 1.

From Figure 1, the use of RSSI for distance estimation is not reasonable since the same RSSI values could correspond to different distances. However, for each individual curve the RSSI values mostly decreased monotonically with increasing distance. There are some abnormal and exceptional situations in the curve which can be attributed to many factors, such as the RF transmission parameters, antenna issues, random noise, propagation path loss and node-level sensing discrepancy. Therefore, because of these factors, the direct use of RSSI values to calculate the physical distance of nodes will result in inaccurate results.

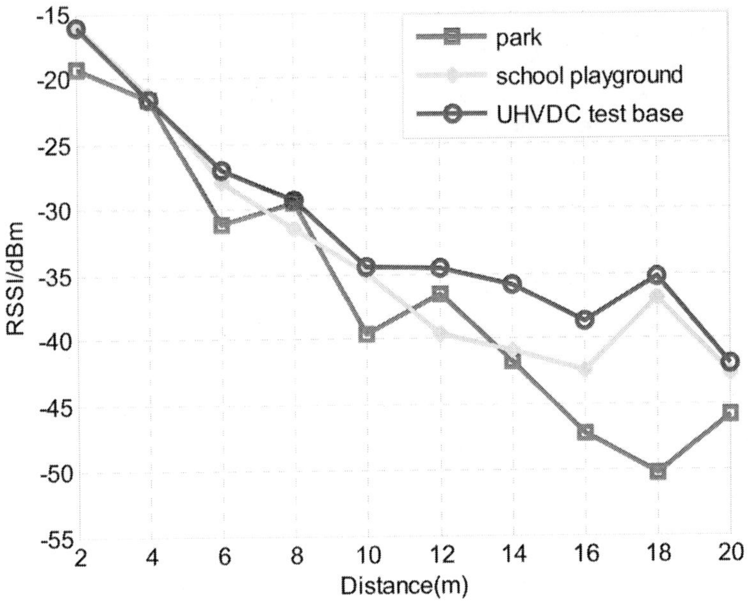


**Figure 1.** The relationship of Received Signal Strength Indicator (RSSI) and physical distance in three types of environments.

(2) Received Signal Strength Indicator (RSSI) ordering and node location permutation

In a linear topology WSN, an ordered node location permutation can be formed by the nodes' physical distance. For any sender node, say *Si*, the sort order of the node location permutation, say *S*, is obtained by the sender node *Si*and the receiver node communication distance from near and far. Another permutation, say *R*, can be obtained by ordering the RSSI value from sender node *Si* to the receiver node in descending order and we define this sorted permutation as RSSI ordering. Ideally, if the RSSI decreases monotonically with the increasing distance, *S* and *R*should be identical. But due to some unexpected factors existing in the process of signal propagation, two permutations *S* and *R* could not be consistent.

As can be seen from Figure 1, the direct use of RSSI to determine the distance between the sender and receiver is not reliable. When sender nodes are far away from the receiver node, it is less accurate to infer their distance by RSSI values. Actually, this is reasonable when the process of signal propagation is considered. Signals spread during propagation and become weaker and weaker. The same noise level can influence weaker signals more easily. As a result, the RSSI becomes less reliable. It is obvious that at different distances from the receiver, the nearest node observes the best signal. In other words, the influence caused by a disturbance will be reduced as the distance between the sender and receiver decreases. This assertion has been verified in the following experiment. In the experiment, four sender nodes are arranged at different distances from the receiver node. All the nodes are in a straight line. The interval between each sender node is set to 1 m, 2 m, and 3 m, respectively. The distance between receiver and the first sender node (Sender 1) is correspondingly set to 1 m, 2 m, and 3 m. The layout of the network is shown in Figure 2a. Each sender node broadcasts 19-bytes test data packets with 18 dB m sending power, and the receiver node records the RSSI upon receiving the packet. The experiment is conducted 100 times at different intervals and the average value of RSSI of each node from different distance is calculated. The experimental result is shown in Figure 2b.

Through the experiment results, the RSSI ordering can be checked as to whether it reveals the true distance relationship among the senders. In Figure 2, the vertical ordinate indicates the accuracy of rank in the RSSI ordering compared with the corresponding rank in node location permutation. The accuracy of the nearest sender node (rank 1 in the node list) is the highest and over 94% in the case of different intervals. This percentage decreases with the increase of rank. The accuracy of the 4th nearest node (rank 4 in the node list) is lower than 32%. Although the percentage decreases with the increase of the interval, the nearest node (rank 1) still obtains a better signal than other nodes. Therefore, the assertion that the nearest node observes the best signal among nodes in different distance from the receiver is verified in the experiment and can be employed to realize node sorting.

## 2.2. Existing Methods as Motivation

The experimental results show that the nearest node usually observes the largest RSSI. Some relative localization methods employing this principle have been proposed [12,13]. The Kcd location algorithm [12] obtains the positon of unknown nodes through the relative localization between the unknown node and the anchor node. The output node permutation can be obtained through the RSSI ordering of unknown nodes. Once the location of an unknown node is known, the node can be upgraded to an anchor node and help to locate the neighboring unknown nodes. This progress is repeated until the localization of as many unknown nodes as possible has been accomplished. However, the Kcd location algorithm has two limitations: (1) it is only suitable for the case where the deployment of the sensor network has a grid or linear topology and where the interval between any two neighboring nodes is regular rather than arbitrary; (2) the Kcd location algorithm upgrades the unknown nodes which have localized their position to an anchor node status, but once an unknown node has failed to localize its positon, the subsequent unknown nodes will all fail to localize their positions.

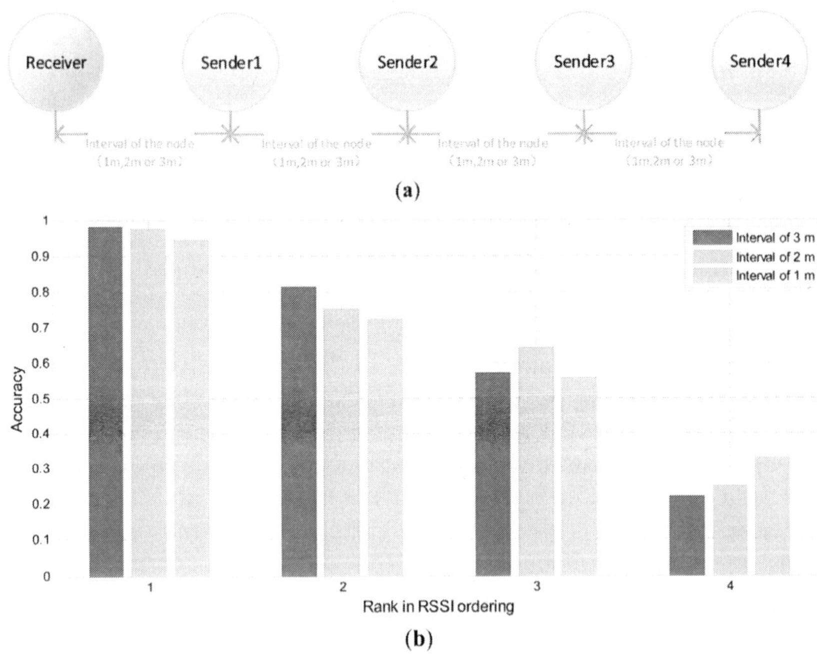


**Figure 2.** The correspondence between the Received Signal Strength Indicator (RSSI) ordering and node location permutation; (**a**) Layout of the network; (**b**) Experimental results.

In [13], a spatial relative localization method without the anchors has been proposed. The method consists of two parts: the first part uses a specific method to find candidate orderings, while the second part verifies the consistency of each candidate ordering and outputs the final results. For each node, it broadcasts the localization command to obtain RSSI and ID from its neighbors. All the neighbors are sorted in descending order according to their observed RSSI. The sorted list is the node neighbor list. The first part of the method is to find the location permutation of nodes with the neighbor list. It iteratively derives the node ordering and starts with each node because there is no information about the initial node. During the iteration, the method deletes the completed one from the node's neighbor list, and prepares for the next iteration until the output of node permutation. If the process cannot proceed, then the method outputs a null. The second part of the method is to verify the consistency of all the reversion of candidate permutations coming from the first part.

Compared with the Kcd location algorithm, the spatial ordering method in [13] can judge whether the output is correct. However, it cannot handle the exception of output node permutation. In practice, three kinds of results will appear when using this method to locate nodes. The first one is all the candidate permutations are wrong, then the method outputs a null; the second result is multiple candidate permutations are simultaneously consistent with the conditions, then the method outputs a null; the third result is only one permutation is consistent with the conditions and the method outputs the right permutation. The spatial ordering method also has several limitations: (1) it does not know the initial node, so it has to iteratively start with every node and derive all the nodes; (2) in the three kinds of results, only the third one can output the right permutation and the others all output a null; (3) it can only judge the correctness of the output permutation, but it cannot handle the exception of the output permutation.

# 3. CORRECT RSSI-BASED RELATIVE-LOCALIZATION (CRBR-LOCALIZATION) ALGORITHM

With respect to the limitations of the localization methods described in [10–13], certain improvements have been carried out and the CRBR-Localization algorithm based on the localization method in [12,13] is achieved. Compared with the localization methods in the literature [12,13], the CRBR-Localization algorithm described in the paper is more applicable in the relative localization of wireless nodes in the electric field measurement environment.

In the localization method of the paper, the nodes that can communicate with node *i* are arranged in descending order according to the return RSSI values, thus obtaining the neighbor list of the node *i*. We select the first three nodes in the neighbor list during the localization, as shown in Figure 3.

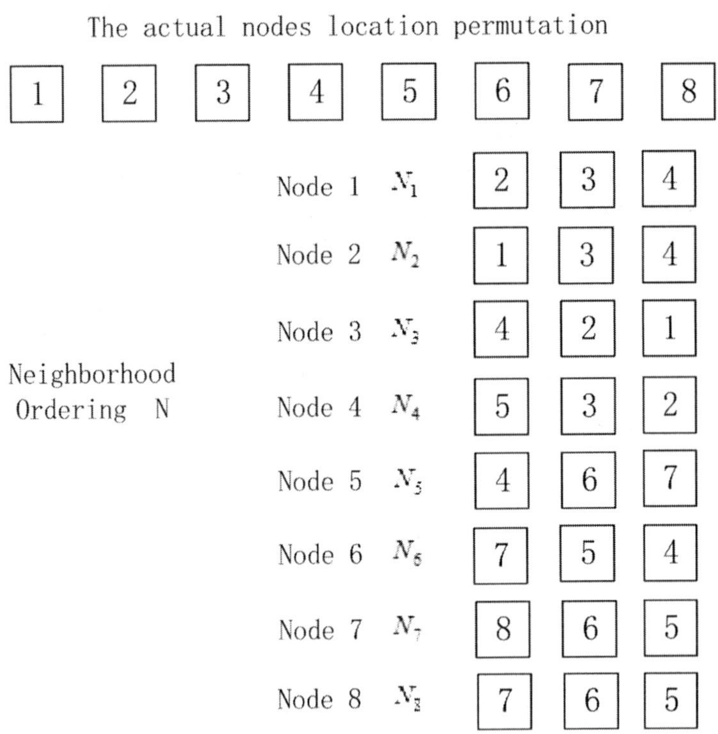


**Figure 3.** Neighbor lists.

In Figure 3, the nodes are arranged in 1–2–3–4–5–6–7–8, which is the actual nodes location permutation. $N_i$ is the neighbor list consisting of three nodes with largest return RSSI values in descending order among nodes communicating with node *i*, while *N* is the set of neighbor list $N_i$ of all nodes.

Before locating the nodes, it is necessary to mark the initial node *A* of the unknown nodes to be sorted. Taking Figure 3 for example, *S* is the output location permutation after the algorithm, and the initial node *S* (1) is node 1.We select the first node (node 2) of the neighbor list of *S* (1) as *S* (2) and delete *S* (1) in the neighbor list of all nodes; select the first node (node 3) of the neighbor list of *S* (2) as *S* (3) and delete *S* (2) in the neighbor list of all nodes. The above process is repeated until the neighbor list of the last node of the ordering is empty, and the output nodes location permutation is *S*. Now we have the first solution and process summarized in Algorithm 1.

**Algorithm 1** Deriving the permutation with initial node

**Input**: all nodes' neighborhood *N*; the initial node *A*.

**Output**: node location permutation *S*.

1. $S(1) = A$; $i = 2$;
2. **while** $N_{S(i-1)}$ is not empty **do**
3. $S(i)$ = the first element of $N_{S(i-1)}$
4. delete $S(i-1)$ in the $N$
5. $i = i + 1$
6. **end while**
7. output the location permutation $S$

## 3.1. Exception Handling

From Figure 2, when the space between nodes is 1 m, over 94% of the RSSI orderings can identify the nearest node correctly, but there are still 6% of RSSI orderings failing to do so. Supposing that the neighbor node with largest RSSI value in the neighbor list of node *i* is not the nearest one to the receiver node during the location, then an inconsistent location permutation will be output. Since there is a small possibility that the above-mentioned anomalous condition may occur, we only analyze the situation where there is only one exceptional node, and the method is not applicable to the situation where several nodes are in an anomalous condition. When there is only one exceptional node among the unknown nodes to be located, two possible results arise: one is that the number of nodes in the output location permutation is less than the actual node number and another one is that the number of nodes in the output location permutation is equal to the actual node number:

(1) The number of nodes in the output location permutation is less than the actual node number.

When there are many nodes to be sorted, there are often two things occurring when the neighbor node with the largest RSSI value is not the one nearest to the receiver node, as shown in Figures 4 and 5. In these figures, the red nodes refer to faulty neighbor nodes while the blue ones are correct neighbor nodes, but the different location of the blue nodes in the neighbor list will cause different output location permutations. In Figure 4, the output location permutation according to Algorithm 1 is 1–2–4–5–6–7–8, where the node number is one less than the actual number, while in Figure 5, the output location permutation according to Algorithm 1 is 1–2–4–3, where the node number is four less than the actual number. In these two cases, the correct location permutation fails to be output due to a certain exceptional node, so a method is desired to correct the abovementioned conditions, thus obtaining the correct output location permutation:

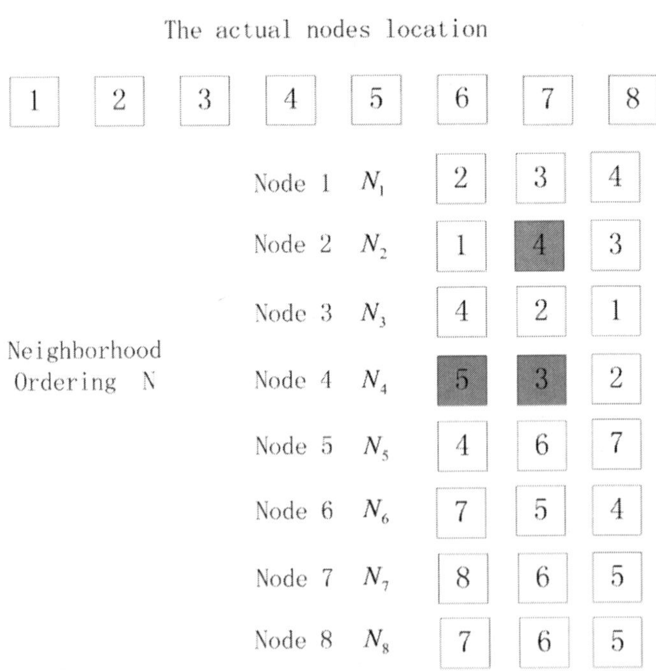


**Figure 4.** Exceptional condition 1.

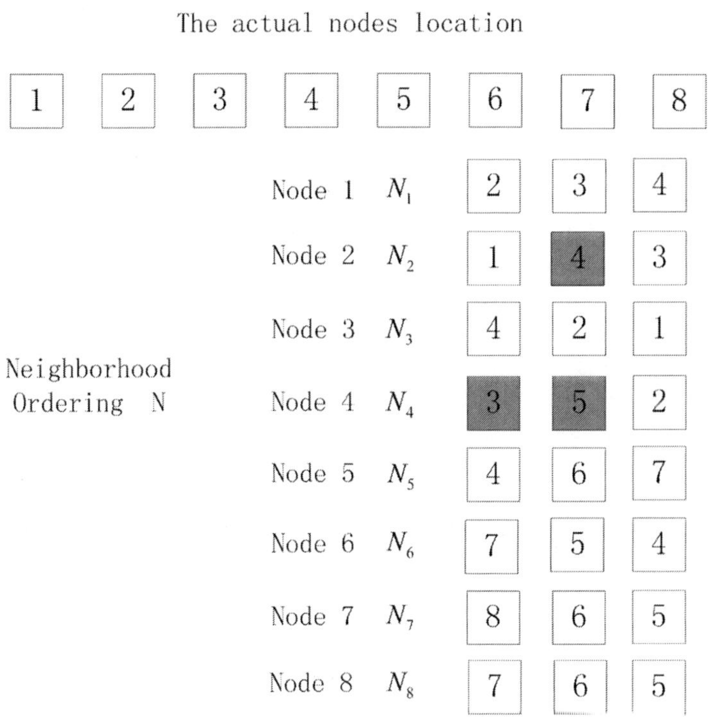


**Figure 5.** Exceptional condition 2.

(a) For the case where the number of nodes in the output location permutation is different from the actual node number by one, the following method can be employed to handle it: choose the node failing to appear in the location permutation, and check whether its first two neighbor nodes adjoin in the ordering; if so, insert the missing node between its two neighbor nodes to get a new ordering, or the sorting fails. Verify the new ordering repeatedly to determine whether it is correct, or the sorting fails. As is shown in Figure 4, Node 3 is missing in the output location permutation, and then check whether the first two nodes in the neighbor list of Node 3 adjoins in the output ordering. The first two nodes in the neighbor list of Node 3 are Node 4 and Node 2, which adjoin in the ordering 1–2–4–5–6–7–8, and then insert Node 3 between Node 4 and Node 2 to get a new ordering 1–2–3–4–5–6–7–8. Select the last node (Node 8) of the new ordering as the initial node and make reverse deduction according to Algorithm 1. If the output ordering is 8–7–6–5–4–3–2–1, it is verified to be successful and output the correct location permutation 1–2–3–4–5–6–7–8. The method to handle such an exception is as shown in Algorithm 2:

(b) For the case where the difference between the number of nodes in the output location permutation and the actual node number is more than one, the following method can be employed to handle it: select the neighbor list of the last but one node in the permutation, and delete the last node of the output location permutation in its neighbor list; if its neighbor list is empty, the output is an empty set and the sorting fails; if there still is a node in its neighbor list, select this node as the next one to continue sorting. When the difference between the node number of the location permutation and the actual node number is more than one, it means more than one node are exceptional, and the sorting fails. When the difference between the node number of the location permutation and the actual node number is one, we handle it according to algorithm 2. Taking Figure 5 for example, the difference between the node number of location permutation and the actual node number is more than one and the location permutation is 1–2–4–3. Select the neighbor list $N_4$ of the last but one node (Node 4) in the ordering and delete Node 3 from $N_4$. Then the first neighbor node in $N_4$ is Node 5 and make deduction according to algorithm 1 to get the order 1–2–4–5–6–7–8. Since the difference between the number of nodes in the location permutation and the actual node number is one, Algorithm 2 can be employed to handle. It can be seen that the location permutation is the same as the one in Figure 4, so Algorithm 2 can be employed to get the correct node location ordering 1–2–3–4–5–6–7–8. The detailed process is shown in Algorithm 3 to handle such an abnormity:

**Algorithm 2** Exception handling algorithm

**Input**: all nodes' neighborhood $N$: error location permutation $S'$.

**Output**: node location permutation $S$.

1. Find the missing node $M$;
2. **If** the first two elements in $N_M$ are neighborhood in the error ordering $S'$ **then**
3. put the missing node $M$ in the $S'$ between the two elements in $N_M$;
4. **else**
5. **break**;
6. $S(1) = S'(n)$, use Alogorithm1 to derive ordering $S$
7. **If** $S$ is the reversion of $S'$ **then**
8. $S = S'$;
9. **else**
10. $S$ = null;
11. **end if** 7. output the location permutation $S$

**Algorithm 3** Exception handling algorithm

**Input**: all nodes' neighborhood N; error location permutation $S'$.
**Output**: node location permutation $S$.
$k$ = the number of $S'$; $n$ = the actual node number
delete the last element of $S'$ in the $N_{S'(k-1)}$
**If** $N_{S'(k-1)}$ is null **then**
    break;
**else**
    $S'(k)$ = the first element of $N_{S'(k-1)}$; delete $S'(k-1)$ in $N$
**end if**
**while** $N_{S'(k)}$ is not empty **do**
    $S'(k+1)$ = the first element of $N_{S'(k-1)}$
    delete $S'(k)$ in $N$
    $k = k + 1$
**end while**
$k'$ = the number of the new ordering $S'$
**If** $k' == n$ **then**
    $S(1) = S'(n)$, use Algorithm 1 to derive the ordering $S$
        **If** $S$ is the reversion of $S'$ **then**
            $S = S'$
        **else**
            $S$ = null
        **end if**
**else if** $n - k' == 1$ **then**
    turn to Algorithm 2
**else**
    $S$ = null
**end if**

(2) The number of nodes in the output location permutation is equal to the actual node number. There are often the following three things occurring when the neighbor node with the largest RSSI value measured is not the one nearest to the receiving node, as is shown in Figure 6, 7 and 8.

In Figure 6, the actual order of nodes is 1–2–3–4–5–6–7–8, but the exceptional neighbor node of Node 5 causes the ordering output according to algorithm 1 to be 1–2–3–4–5–7–8–6. When in such a case, the reverse ordering of the output location permutation, $S(n)$ $S(n-1)$ ..., $S(1)$, can be verified to determine whether the output location permutation is correct, that is to take $S(n)$ as the initial node and make ordering deduction according to algorithm 1 to get the location permutation, which is called the reverse verification permutation S1. If the output reverse verification permutation is identical to the reverse ordering of permutation S, the sorting is correct, and if not, the sorting is faulty. In Figure 6, permutation $S$ is 1–2–3–4–5–7–8–6, and the reverse verification ordering deducted with Node 6 as the initial node is 6–7–8–5–4–3–2–1, which is not consistent with the expected reverse verification permutation, 6–8–7–5–4–3–2–1, so the node sorting is faulty.

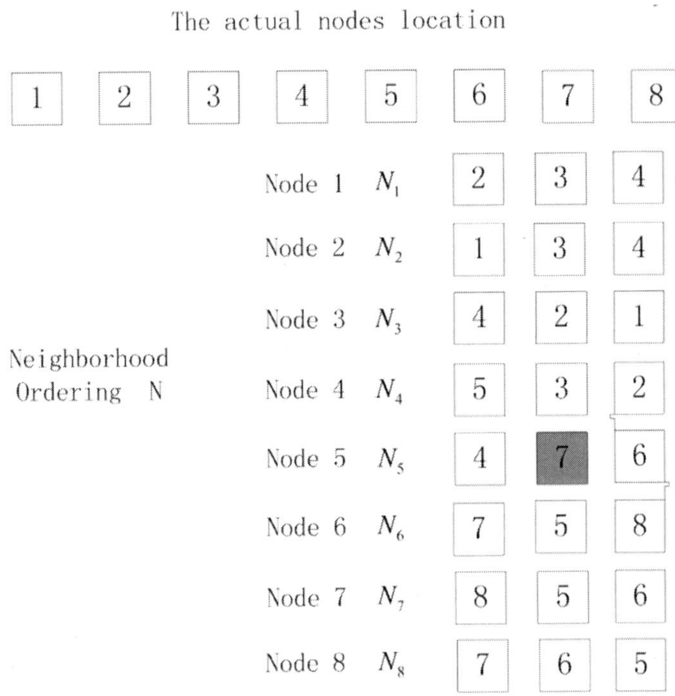


**Figure 6.** Exceptional condition 3.

In Figure 7, the actual order of nodes is 1–2–3–4–5–6–7–8, and although the neighbor node of Node 7 is exceptional, the ordering output according to Algorithm 1 is 1–2–3–4–5–6–7–8. However, when taking Node 8 as the start node, the reverse verification ordering output according to Algorithm 1 is 8–7–5–4–3–2–1, which is not consistent with the expected reverse verification ordering, 8–7–6–5–4–3–2–1.

In Figure 8, the actual order of nodes is consistent with that in Figure 7, and its ordering permutation according to Algorithm 1 is still 1–2–3–4–5–6–7–8, but its reverse verification permutation according to Algorithm 1 is 8–7–5–6, which is not consistent with the expected reverse verification ordering, 8–7–6–5–4–3–2–1.

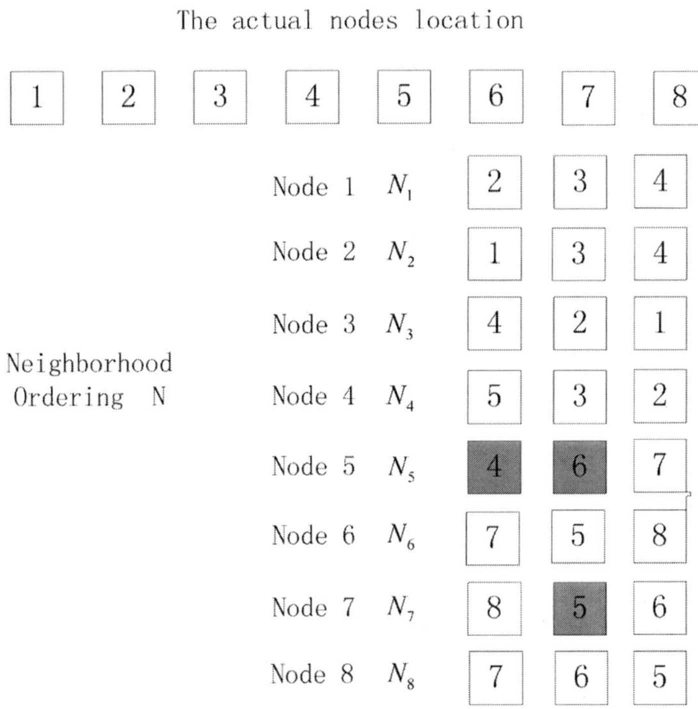


**Figure 7.** Exceptional condition 4.

The three cases mentioned above can be handled according to the following algorithm. When permutation $S$ is deducted according to Algorithm 1,whether the output location ordering is correct can be determined by comparing whether the reverse verification permutation $S1$ is consistent with the reverse ordering of permutation $S$. If the node number of permutation $S1$ is identical to that of permutation $S$, and it is the reverse ordering of permutation $S$, the sorting is correct; if the difference between the node numbers of permutation $S1$ and permutation $S$ is one, it can be corrected according to algorithm 2 using $S1$ as input; if the difference between the node numbers of permutation $S1$ and permutation $S$ is more than one, it can be corrected according to Algorithm 3 using $S1$ as input; if it cannot be corrected, the sorting fails. When in the three above-mentioned cases, only Figure 6 fails to output correct ordering while in the other two cases, relevant algorithms can be used to get correct node location ordering. The detailed process is shown in Algorithm 4.

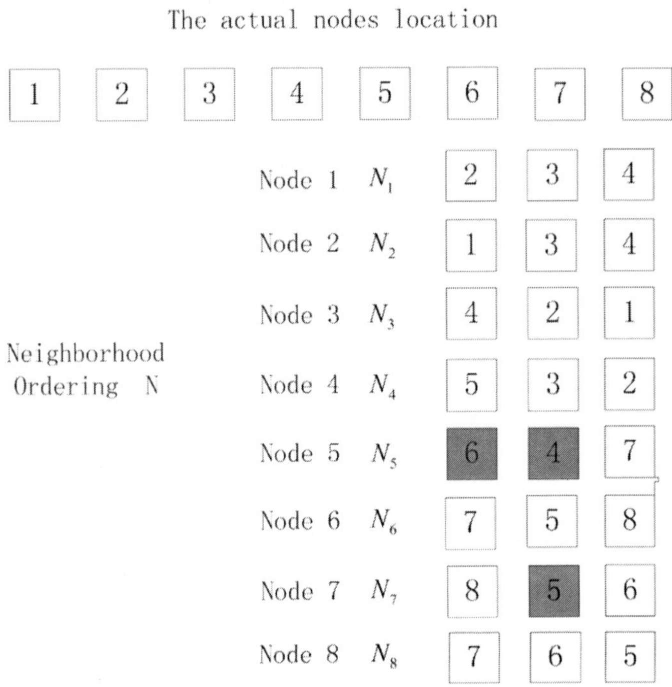


**Figure 8.** Exceptional condition 5.

**Algorithm 4** Exception handling algorithm

**Input**: all nodes' neighborhood $N$: verified permutation $S$

**Output**: node location permutation $S$.

1. use Alogorithm1 to derive the reverse verification permutation $S1$
2. $k$ = the number of ordering $S1$; $n$ = the actual node number
3. **If** $k == n$ **then**
4.     **If** $S1$ is reversion of $S$ **then**
5.     $S = S$
6.     **else**
7.     $S$ = null
8.     **end if**
9. **else if** $n - k == 1$ **then**
10.     turn to Algorithm 2 *using S1 as input*
11. **else**
12.     turn to Algorithm 3 *using S1 as input*
13. **end if**

### 3.2. Realization of Correct RSSI-Based Relative-Localization (CRBR-Localization) Algorithm

The CRBR-Localization algorithm combines the two methods in [12,13], and applies a correction algorithm with respect to their respective shortcomings, which make it possible to handle the exception where one node fails. However, CRBR-Localization algorithm is only capable of correcting the cases in Figures 4, 5, 6, 7 and 8, and for the case where several nodes fail, the algorithm output ordering is empty. The realization of the CRBR-Localization algorithm is shown in Algorithm 5:

---

**Algorithm 5** CRBR-Localization Algorithm

**Input**: all nodes' neighborhood $N$; the initial node $A$; the actual node number $n$.
**Output**: node location permutation $S$.
1. use Algorithm 1 to derive the permutation $S$
2. $k$ = the number of ordering $S$
3. **If** $k == n$ **then**
4. turn to Algorithm 4
5. **else if** $n - k == 1$
6. turn to Algorithm 2
7. **else**
8. turn to Algorithm 3
9. **end if**

---

## 4. SYSTEM IMPLEMENTATION AND EVALUATION

To evaluate the CRBR-Localization algorithm, three indexes are chosen: (1) percentage of output, which is the ratio of the occurrences that the output permutation is a non-empty set after sorting to the experiment occurrences; (2) output accuracy, which is the ratio of the occurrences that the output permutation is a right sequence to the occurrences the output permutation is a nonempty set after sorting; (3) accuracy, the ratio of the occurrences that the output permutation is a right one after sorting to the experiment occurrences. The proposed three indexes can be defined as Equations (1)–(4):

$$R_{po} = \frac{N_{non}}{N_{all}} \tag{1}$$

$$R_{oa} = \frac{N_{acu}}{N_{non}} \tag{2}$$

$$R_{acu} = \frac{N_{acu}}{N_{all}} \tag{3}$$

$$R_{acu} = R_{po} \times R_{oa} \tag{4}$$

where $R_{po}$ is percentage of output, $R_{oa}$ is output accuracy, $R_{acu}$ is accuracy, $N_{non}$ is the occurrences that the output permutation is a nonempty set after sorting, $R_{acu}$ is the occurrences that the output permutation is a right sequence after sorting, $R_{all}$ is the total experiment occurrences. From the definitions of the three indexes, we can know that accuracy ($R_{acu}$) equals to the arithmetic product of the percentage of output ($R_{po}$) and the output accuracy ($R_{oa}$). The ideal algorithm has not only a high percentage of output but also high output accuracy. Meanwhile, the index of accuracy could be a comprehensive evaluation of the algorithm's percentage of output and output accuracy.

## 4.1. Experimental System

To validate and verify the proposed localization algorithm, a double circuit experimental line on the same tower was erected at the HVDC test base of China Electric Power Research Institute in Beijing. Figure 9 shows our hardware system for the localization.

There are two kinds of nodes which are wireless nodes (unknown nodes) and a sink node in the system. The sink node is connected to the computer through the USB bus. The wireless nodes which are arranged in a line perpendicular to the transmission lines are the subject of the localization. The sink node broadcasts the localization command message to the wireless nodes. Upon receiving the message, the wireless nodes will communicate with their neighbor nodes and obtain the neighbor list. In the same way, all the wireless nodes in turn get their neighbor lists and then transmit them to the sink node. As a result, the computer connected with the sink node executes the CRBR-Localization algorithm. The detailed workflow of the CRBR-Localization algorithm is shown in Figure 10. The sink node has two main functions. The first one is to broadcast the localization commands and another one is to receive all the unknown nodes' neighbor lists used for the CRBR Localization algorithm. There is a high-performance CPU and enough memory in each unknown node, so the neighbor list of each unknown node is created by the unknown node itself and then sent to the sink node. Therefore in the experimental system, only one node which is connected with the computer through the USB interface is the sink node. The sink node could be outside or at one end of the network of unknown nodes. It is convenient for the operator to always take as the sink node the one with the connected computer.

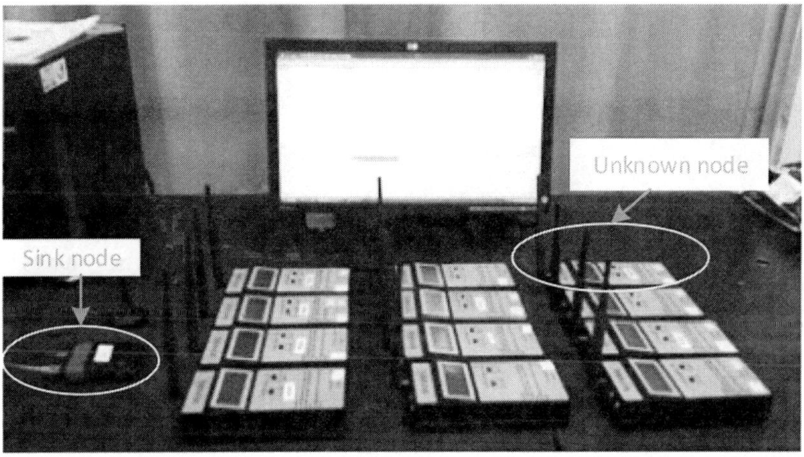


**Figure 9.** Experiment system at the High Voltage Direct Current (HVDC) test base of the China Electric Power Research Institute.

A block diagram of the proposed wireless node is presented in Figure 11. The wireless nodes' primary function is to acquire and process the output analog signals from the electric field sensors and send the digital data to the sink node. The core of the wireless node is an STM32F embedded microcontroller. The wireless communication system selected for the sensor node is the 2.4 GHz Xbee Pro Radio Frequency (RF) module based on the IEEE 802.15.4 protocol from Digi™(Minnetonka, MN, USA).

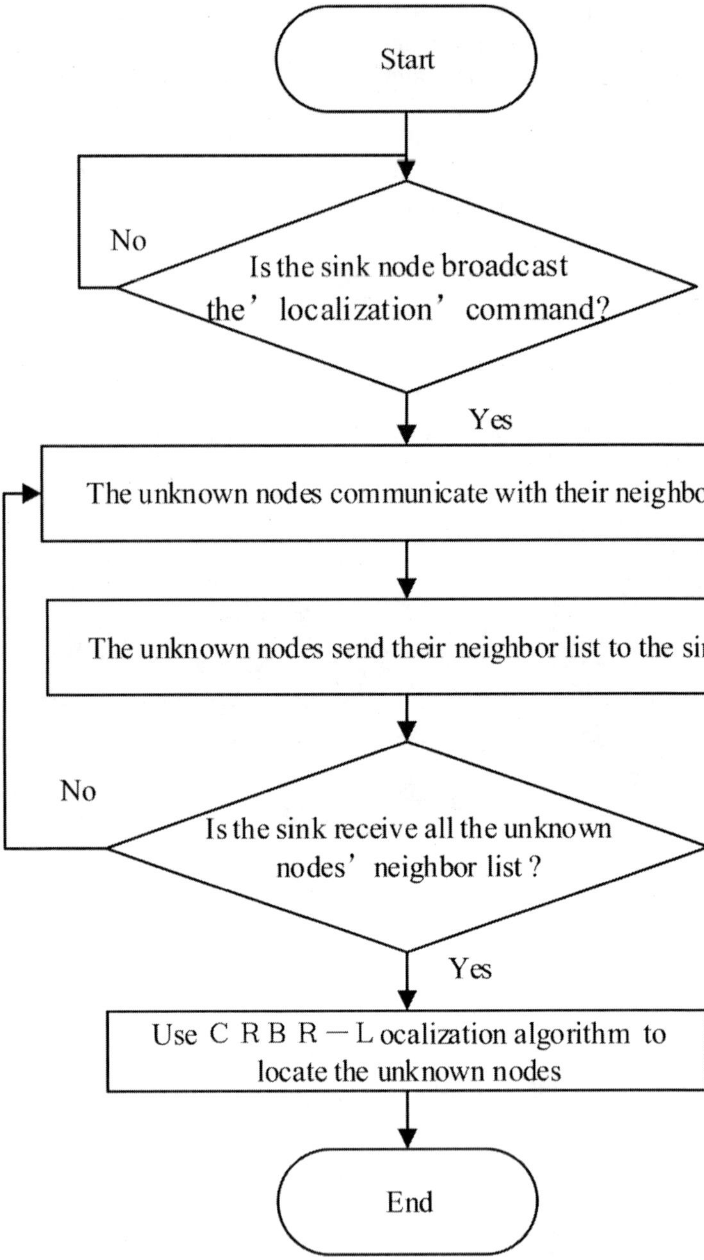


**Figure 10.** Workflow of the Correct RSSI-Based Relative-Localization (CRBR-Localization) algorithm.

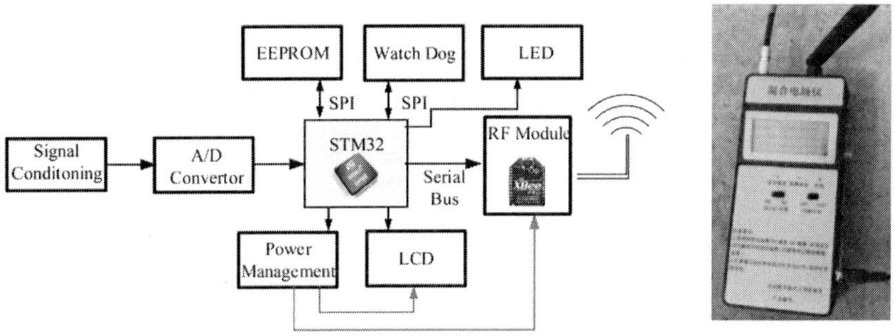


**Figure 11.** Structure and picture of a wireless node.

## 4.2. Node Number Experiments

In the electric field measurement environment, since the transmission line spacing is different, different numbers of sensor nodes are needed. In order to verify the validity of the algorithm, we did sorting experiments using different node numbers. In these experiments, the numbers of unknown nodes to be positioned are 5, 10, 15, and 20, respectively. The nodes are arranged in a line perpendicular to the transmission lines, and the interval between nodes is 2.5 m. The RSSI values that are returned by the unknown nodes in each sequence are sampled five times and averaged in order to use the average for sorting. Three hundred sequences are obtained with each sorting, and the sorting accuracy is counted and analyzed. The result of the experiment is shown in Figure 12.

Since the output accuracies of the experiments are all 100%, the accuracy equals the percentage of output, thus only accuracy is used as the evaluation criterion. From Figure 12 we can see that when sorting five nodes, the sorting accuracies of nodes before and after exception handling are both 98.4%, thus when the number of nodes is small, exception handling has an insignificant influence on the result. Node sorting accuracy before exception handling falls dramatically with the increase of the number of sorting nodes. When the number of sorting nodes is 10, the accuracy is 94.4%, and when the number of sorting nodes is 15, the accuracy is 91.25%. When the number of sorting nodes is 20, the accuracy is only 86.6%. Node sorting accuracy after exception handling falls slightly with the increase of the number of sorting nodes. When the number of sorting nodes is 10 the accuracy is 98.1%, and when the number is 15, the accuracy is 97.5%. When the number of sorting nodes is 20, the accuracy is still above 96.3%, so when the number of nodes in the experiment is large, the sorting accuracy can be improved markedly through exception handling by the CRBR-Localization algorithm.

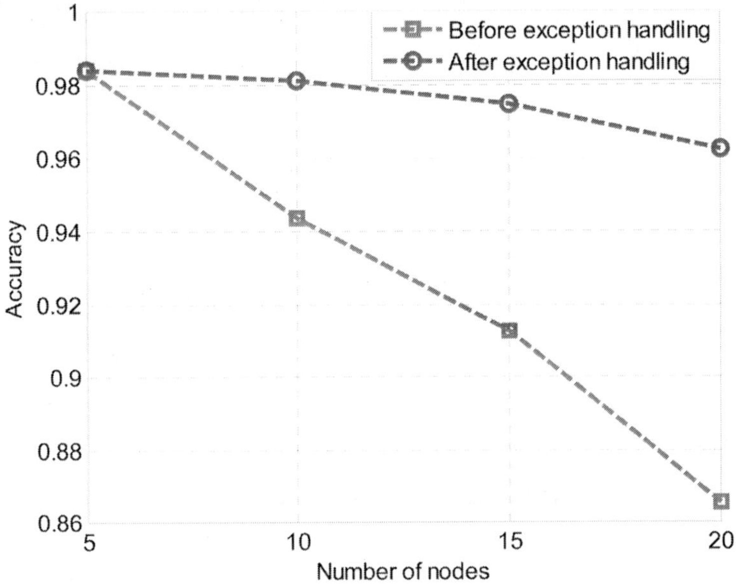


**Figure12.** Experimental results.

## 4.3. Experiment in Four Deployments

Theoretically, when placing sensor nodes, every node should be arranged in a line perpendicular to the transmission lines. However, since there exist personal deployment errors, not all the nodes are in a straight line, so the adaptability of the CRBR-Localization algorithm to different node deployment structures needs to be verified.

In the experiment, we sequence 10 unknown wireless nodes in four different deployment structures. The RSSI values that are returned by the unknown nodes in each sorting are sampled five times and averaged for further sorting. Three hundred sequences are obtained with each structure. The results are obtained to compare with the Spatial Ordering method presented in [13] and the Kcd location algorithm presented in [12].

### 4.3.1. Linear Homogeneous Deployment

In this experiment, the interval between neighboring nodes is 3 m. The nodes are arranged in a line perpendicular to the transmission line, as shown in Figure 13. The distance between the first node and the last node is 27 m, and they can communicate normally. The result is shown in Figure 14.

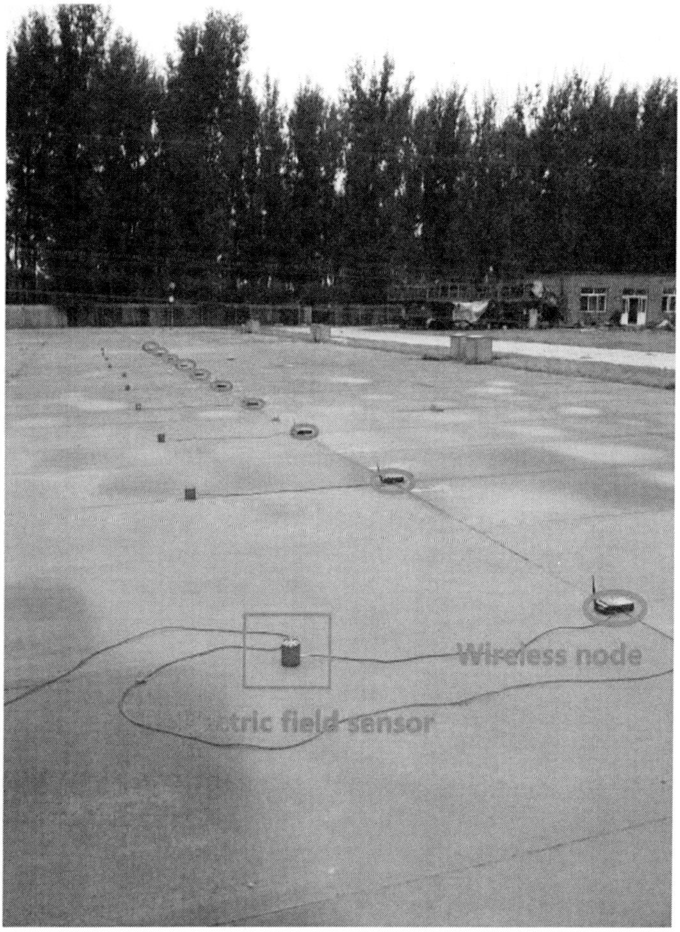


**Figure 13.** Linear homogeneous deployment structure.

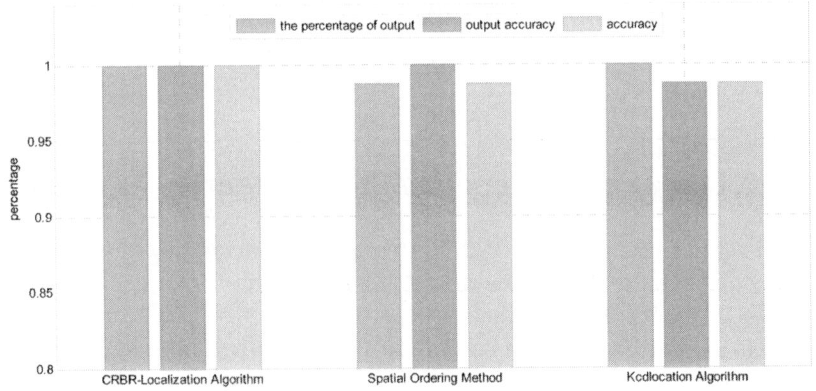


**Figure 14.** Localization results comparison by using different algorithms in the linear homogeneous deployment.

From Figure 14 we can see that the percentage of output and the output accuracy of the CRBR-Localization algorithm are both 100%, higher than the values of the other two algorithms. Since the Spatial Ordering method doesn't process exceptional conditions and may have several candidate orderings, its percentage of output is lower than that of the CRBR-Localization algorithm and Kcd location algorithm. Because the Kcd location algorithm can't judge the correctness of its output sequences, although the percentage of output is 100%, the output accuracy is still lower than those of the CRBR-Localization algorithm and Spatial Ordering method. Comprehensively comparing the three algorithms, we can see that the three indexes of the CRBR-Localization algorithm are all higher than those of the other two algorithms, and it has higher localization accuracy.

### 4.3.2. Linear Inhomogeneous Deployment

In this experiment, the intervals between neighboring nodes are not the same. The maximum interval is 4 m, while the minimum is 1.5 m. The nodes are arranged in a line perpendicular to the transmission lines, as shown in Figure 15. The first node and the last node can communicate normally, as shown in Figure 16.

From Figure 16, Kcd location algorithm has the highest percentage of output, and the CRBR-Localization algorithm is in the middle, while the

Spatial Ordering method has the lowest, which is less than 90%. Although the Kcd location algorithm has the highest percentage of output, its accuracy and output accuracy are both lower than 90%, apparently lower than those of the CRBR-Localization algorithm.

**Figure 15.** Linear inhomogeneous deployment structure.

The percentage of output of the CRBR-Localization algorithm is higher than 99%, and its accuracy is higher than 96%, both of which are higher than Kcd location algorithm and Spatial Ordering method. Under linear inhomogeneous deployment conditions, the percentage of output of CRBR-Localization algorithm and Spatial Ordering method are both lower than under a linear homogeneous alignment. This is because when the nodes are placed homogeneously, there may be a situation that some nodes are placed closely and others are placed loosely. When the nodes are close, the distance between nodes is small, so the RSSI values measured are close. Since the RSSI values of each node are close, elements like electromagnetic components, topographic relief and surface vegetation will interfere with the radio signal, which will have a greater influence on obtaining the RSSI accurately, whereas the CRBR-Localization algorithm and the Spatial Ordering method can judge the correctness of output location permutation, and when the output permutation is judged as a wrong one, then output will not be allowed. In addition to that, the CRBR-Localization algorithm can not only judge the correctness of output permutations, but also further correct the wrong permutations, so it has a higher percentage of output and accuracy than the Spatial Ordering method.

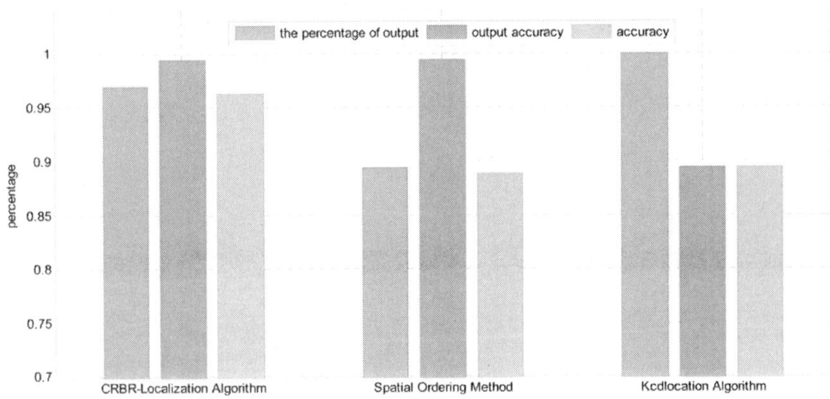


**Figure 16.** Localization results comparison by using different algorithms in the linear inhomogeneous deployment.

### 4.3.3. Cross Deployment

In this experiment, each node is first arranged in a line with a 3 m interval. Then we make each node move 0.5 m to the left and right, respectively, as is shown in Figure 17. From Figure 18, the CRBR-Localization algorithm has an output accuracy of 100%, and has a higher accuracy than the Kcd location algorithm and Spatial Ordering method. Though the Kcd location algorithm has a high percentage of output, it can't judge the correctness of the output location permutations, so its percentage of output and accuracy are both lower than those of the CRBR-Localization algorithm. Comprehensively comparing the three algorithms, we can see that in cross deployment structure scenarios, the overall performance of the CRBR-Localization algorithm is better than that of the other two algorithms.

**Figure 17.** Cross deployment structure.

## 4.3.4. Random Deployment

In actual measurement, since there exist personal deployment errors, so not all the nodes are arranged according to the ideal structure (linear and homogeneous). Therefore, an approximate random arrangement is used to simulate the situations in an actual measurement environment. Under this deployment, the nodes are arranged randomly in a rectangle 30 m long and 1 m wide, and the minimum interval of adjacent nodes is 1.8 m, while the maximum is 3.6 m, as shown in Figure 19.

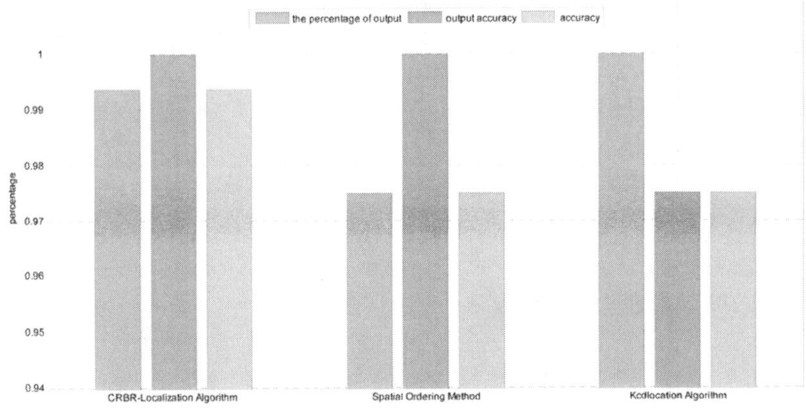


**Figure 18.** Localization results comparison by using different algorithms in the cross deployment.

Under random deployment, the intervals between nodes are not the same, and not all the nodes are in the same line, but they are randomly arranged in a rectangle area. Under this circumstance, the RSSI is sensitive to the interference factors in the environment. From Figure 20, we can see that the CRBR-Localization algorithm has accuracy and percentage of output higher than 97%, and its output accuracy is 100%. Both its output accuracy and its percentage of output are higher than those of the Kcd location algorithm and Spatial Ordering method. Though the Spatial Ordering method has a higher percentage of output, its accuracy and output accuracy are both lower than 94%, which is lower than the CRBR-Localization algorithm. Comprehensively comparing the three algorithms, we can see that under random arrangement conditions, the overall performance of the CRBR-Localization algorithm is better than that of the other two algorithms. In the four different localization experiments, the accuracy and percentage of output of the CRBR-Localization algorithm are apparently higher than those of the Kcd Location algorithm and the Spatial Ordering method.

**Figure 19.** Random deployment.

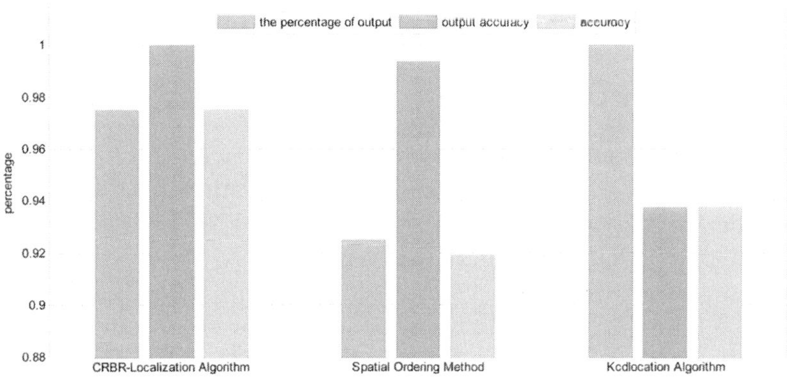


**Figure 20.** Localization results comparison by using different algorithms in the random deployment.

Four different deployment structures, the accuracies of the CRBR-Localization algorithm are all above 96%. Taking the three performance indexes into consideration, we can see that compared with the other two algorithms, the CRBR-Localization algorithm has a better overall performance, and has a better adaptability of different applied environments of electric field measurement.

## 5. CONCLUSIONS

In this paper, a RSSI-based algorithm is proposed for achieving the relative localization of one-dimensional linear topology wireless networks under HVDC transmission lines. Starting from the observation that the closest node always has the highest RSSI, the algorithm utilizes unknown nodes' neighbor list based on the RSSI values to determine the relative locations of nodes. A detailed interpretation has been made around the solutions of the exceptional situations for the output permutation which can effectively improve the localization accuracy. The performance of this algorithm in real circumstances has been evaluated through several experiments with different numbers of nodes and different deployments at the China State Grid HVDC test base. Results show that in the experiments with different number of nodes, the accuracy of the CRBR-Localization algorithm is above 96.2% and in the experiments with different deployments the accuracy of the CRBR-Localization algorithm, which is over 96%, is higher than that of the other two algorithms.

## ACKNOWLEDGMENTS

The authors would like to thank all the reviewers for their helpful comments. The research was supported by National Natural Science Foundation of China (61273165, 51207005) and Basic Research Foundation of Beihang University (YWF-14-ZDHXY-13).

## AUTHOR CONTRIBUTIONS

All authors have made significantly contributions to the paper. Cui Yong carried out the literature survey and wrote the paper section about the relative localization methods for the measurement of the electric field under HVDC transmission lines. Yuan Haiwen proposed the concept of relative localization in the WSN. The experimental system was developed and operated by Hu Xuemin under the supervision of Wang Qiusheng. Song Xiao also contributed to the data collection and modification of the manuscript. Zhao Luxing provided experimental conditions and gave valuable suggestions for improvement in the China State Grid HVDC test base.

## REFERENCES

1. Kasdi, A.; Zebboudj, Y.; Yala, H. Calculation and measurement of electric field under HVDC transmission lines. *Eur. Phys. J.-Appl. Phys.* **2007**, *37*, 323–329.

2. Luo, Z.; Cui, X.; Zhang, W.; Lu, J. Calculation of the 3-D ionized field under HVDC transmission lines. *IEEE Trans. Magn.* **2011**, *47*, 1406–1409.

3. Cui, Y.; Lv, J.; Yuan, H.; Zhao, L.; Liu, Y.; Yang, H. Development of a wireless sensor network for distributed measurement of total electric field under HVDC transmission lines. *Int. J. Distrib. Sens Netw.* **2014**, *2014*.

4. Yang, Y.; Lu, J.; Lei, Y. A calculation method for the electric field under double-circuit hvdc transmission lines. *IEEE Trans. Power Deliver* **2008**, *23*, 1736–1742.

5. Zhang, Q.; Foh, C.H.; Seet, B.-C.; Fong, A. Variable elasticity spring-relaxation: Improving the accuracy of localization for WSNs with unknown path loss exponent. *Pers. and Ubiquit. Comput.* **2012**, *16*, 929–941.

6. Karalar, T.C.; Rabaey, J. An RF ToF Based Ranging Implementation for Sensor Networks. Proceedings of the 2006 IEEE International Conference on Communications (ICC), Istanbul, Turkey, 11–15 June 2006; pp. 3347–3352.

7. Cheng, X.; Thaeler, A.; Xue, G.; Chen, D. A Time-Based Positioning Scheme for Outdoor Wireless Sensor Networks. Proceedings of the 2004 Twenty-Third AnnualJoint Conference of the IEEE Computer and Communications Societies (INFOCOM), 7–11 March 2004, Hong Kong, China; pp. 2685–2696.

8. Sayed, A.H.; Tarighat, A.; Khajehnouri, N. Network-based wireless location: Challenges faced in developing techniques for accurate wireless location information. *IEEE Signal Proc. Mag.* **2005**, *22*, 24–40.

9. Jiang, J.-A.; Zheng, X.-Y.; Chen, Y.-F.; Wang, C.-H.; Chen, P.-T.; Chuang, C.-L.; Chen, C.-P. A distributed RSS-based localization using a dynamic circle expanding mechanism. *IEEE Sens. J.* **2013**, *13*, 3754–3766.

10. Zhong, Z.; He, T. Achieving range-free localization beyond connectivity. Proceedings of the 7th ACM Conference on Embedded Networked Sensor Systems, Berkeley, CA, USA, 4–6 November 2009; pp. 281–294.

11. Yang, X.; Kong, Q.; Xie, X. One-dimensional localization algorithm based on signal strength ratio. *Int. J. Distrib. Sens. Netw.* **2009**, *5*, 79–79.

12. Fang, Z.; Zhao, Z.; Cui, X.; Geng, D.; Du, L.; Pang, C. Localization in wireless sensor networks with known coordinate database. *EURASIP J. Wirel. Comm. Netw.* **2010**, *2010*, 2.

13. Zhu, X.; Wu, X.; Chen, G. Relative localization for wireless sensor networks with linear topology. *Comput. Commun.* **2013**, *36*, 1581–1591.

14. Karalar, T.C.; Yamashita, S.; Sheets, M.; Rabaey, J. A Low Power Localization Architecture and System for Wireless Sensor Networks. Proceedings of the 2004 IEEE Workshop on Signal Processing Systems (SIPS), Austin, New Zealand, 13–15 October 2004; pp. 89–94.

15. Hightower, J.; Want, R.; Borriello, G. *Spoton: An Indoor 3D Location Sensing Technology Based on RF Signal Strength*; UW CSE 00-02-02; University of Washington, Department of Computer Science and Engineering: Seattle, WA, USA; 18; February; 2000.

16. Niculescu, D.; Nath, B. DV based positioning in ad hoc networks. *Telecommun. Syst.* **2003**, *22*, 267–280.

17. He, T.; Huang, C.; Blum, B.M.; Stankovic, J.A.; Abdelzaher, T. Range-Free Localization Schemes for Large Scale Sensor Networks. Proceedings of the 9th Annual International Conference on Mobile Computing and Networking, San Diego, CA, USA, 14–19 September 2003; pp. 81–95.

7

# AN ECO-HYDROLOGY WIRELESS SENSOR DEMONSTRATION NETWORK IN HIGH-ALTITUDE AND ALPINE ENVIRONMENT IN THE HEIHE RIVER BASIN OF CHINA

**Yaonan Zhang[1,2], Lihui Luo[1,2], Jiuyuan Huo[1,2], Wenpin Zhu[1,2]**

[1]Cold and Arid Regions Environmental and Engineering Research Institute, Chinese Academy of Sciences, Lanzhou, China

[2]Gansu High-Performance Grid Computing Center, Lanzhou, China

## ABSTRACT

The emerging of Wireless Sensor Network (WSN) technology has brought new opportunities for data intensive observation in eco-hydrological research and environment monitoring. To make up for the insufficiencies of space observation data for models in eco-hydrology processes research of inland river basin which constrain the construction of accurate modeling and simulation, a sub-basin of the Heihe River Basin named Hulugou has been selected to perform the deployment and research for wireless sensor network from the perspective of eco-hydrology. In this paper, we discussed the system framework, the demonstration research area, the deployment of sensor nodes and the network topology of wireless sensor network. Finally, the study has summarized existing problems and experiences of wireless sensor network in these eco-hydrological applications and the available researches in hard environment. In the future, we will deploy a large number of sensor nodes in the long-term fine-grained eco-hydrological elements observing environment to form lots of homogeneous or heterogeneous wireless sensor networks.

## KEYWORDS

Wireless Sensor Network; Eco-Hydrology; Intensive Observation; Wireless Transmission; Environmental Monitoring

## 1. INTRODUCTION

At present, there are many problems existing in eco-hydrology processes researches of river basin which constrain the construction of accurate modeling and simulation [1]. One of the key problems is the data needed for modeling, especially the space observation data is insufficient. Thus, preparing continuous, systematic, reliable, and intensive data for model construction and simulation is important for river basin ecology-hydrology research. The lack of data on the true variability of the meteorological data in space and time hinders the model's ability to accurately represent short term or localized responses in river flow.

Current observation methods in field environment, such as the traditional fixed-point observation stations are difficult for large-scale deployment. High cost, high risk, intensive labor works and other problems for highdensity deployment in river basin. Therefore, a new method for observation is needed to realize intensive data observation to make up for the sparse distribution of the traditional fixed-point observation stations and to meet the requirements of eco-hydrological model building and simulation.

With the rapid development and increasingly mature of Micro Electro Mechanical systems (MEMS) technology, computer technology, wireless communication and sensor technology, the emerging of Wireless Sensor Network (WSN) technology has brought new opportunities for data intensive observation of eco-hydrological research [2]. From the 21st century, WSN technology has attracted great concern in industrial, military and academic fields and expended application prospects and potential application values in the ecological environment research.

WSN integrates environment awareness, data processing, wireless communication and energy supply into a sensor node, which forms the wireless sensor network with self-organizing approach to realize realtime sensing, monitoring, collecting and processing information of various monitoring objects. It comes into being an intensive observation technique of low-cost and large area deployment. In February 2003, the "Technology Review" magazine of U.S. elected the top ten emerging technologies which have imposed profound impact on human life and WSN is listed as one of them. WSN was taken for as one of the four pillar industries in 21st century high-tech field in technical comments of the U.S. "Business Weekly" magazine in August 2003. Similar to the satellite remote sensing technology produced in the 20th century, wireless sensor network will bring out revolutionary change in all aspects of the earth system and environmental science [3].

Environmental monitoring has become an important application field of WSN and has undertaken a number of typical applications at present. Environmental sensor networks (ESNs) provide new opportunities for improving our understanding of the environment. W.R. Philip, et al. integrated complex sensors, such as networked digital cameras and microphones into sensor networks for hierarchical methods of sensing that promise a further understanding of our ecological systems by revealing previously unobservable phenomena [4]. S.S.O. Burgess, et al. described the current state of wireless environmental monitoring from an end-user point of view and report their initial attempts to adapt the cutting-edge technology to the realm of forest research [5]. A novel environmental monitoring system with a focus on overall system architecture for seamless integration of wired and wireless sensors for long-term, remote, and near-realtime monitoring was presented in [6]. J.V. Borre, et al. has tried to integrate remote sensing technology in their habitat monitoring project—Natura 2000 [7]. To obtain water and salinity mass balances for a 60,000 ha tract of seasonally managed wetlands in the San Joaquin River Basin of California, N.W.T. Quinn, et al. described the use of a number of state-of-the-art sensor technologies [8]. And X. Zhu describes an online water quality monitoring system for intensive fish culture in China, which combined web-server-embedded technology with mobile telecommunication technology [9]. The GENESIS project provides an efficient Web-based software solution that can be used by all kind of actors involved in environment management and health services in Europe [10]. C. Youn, et al. presents the TEAM data management, access, and analysis system that provides end-to-end solutions for sensor-based data and field observations [11]. In [12], a Wireless Sensor Network for Seismic-Observation-Environment

Surveillance has been designed and described. need to create these components, incorporating the applicable criteria that follow.

# 2. RESEARCH AREA AND RESEARCH OBJECTS

## 2.1. Research Area

Heihe River originates on the northern Gansu side of the Qilian Shan mountains, then flows through Hexi Corridor of Gansu Province, ends in Juyanhai and the West Juyanhai of Ejinaqi in the territory of Inner Mongolia Autonomous Region. Its total length is 821 km. Heihe River Basin covering an area of approximately 142,900 $Km^2$ (96°42'E - 102°04'E, 39°45'N - 42°40'N) is the second largest inland river basin of Northwest China. According to altitude and natural geographical characteristics, it can be divided into three types of landscapes including upstream Qilian Mountains, midstream corridor plain and downstream Alxa Plateau. The upper stream altitude ranging between 2500 to 5000 m is the Qilian Mountains located in the northern margin of QinghaiTibet Plateau and Climatic characterized by more precipitation, less evaporation, low temperature, cold and wet, and the Tuanjie Peak with about 5826 m high is the birthplace and source region of Heihe river. Zhengyi gorge is the boundary between the midstream and downstream.

As shown in **Figure 1**, the Hulugou sub-basin (99°50'E - 99°54'E, 38°12'N - 38°17'N) is a tributary of the Heihe River Basin which located in Qilian county of Qinghai province. The drainage area is about 23.1 $km^2$ with the elevation ranging from 2960 m and 4820 m, a typical continental alpine mountain climate which has an annual total precipitation between 400 mm to 600 mm and temperature ranges from –30°C to 30°C. The climate is variable and changeable from day to day. Weather is generally cold and windy with frequent cloud and rain. Only five months throughout the year is snow-free. Due to extremely adverse weather conditions, it have created a local fragile ecology, sparse vegetation, and mostly cold, drought tolerance and short-lived alpine plant [13]. However, the alpine cold desert zone plays a water tower role in hydrological cycle as one of the birthplaces of main rivers in cold regions, which is sensitive to climate warming. Environmental parameter monitoring in alpine cold desert zone can provide data support for feedback land surface branch of high-precision multi-scale atmospheric circulation patterns, and runoff out of the mountains and water balance prediction under the condition of climate change.

## 2.2. Research Quadrat

To study the interactions between atmosphere and surface of Hulugou steppe meadow in the upper streams of Heihe River Basin, a 400 × 400 $m^2$ prairie meadow at the entrance of a small valley was selected as a demonstration research area of wireless sensor network. The area was divided into 100 × 100 $m^2$ grid, the appropriate sensor probes of air temperature, air humidity, solar radiation, soil temperature and soil humidity were deployed in the corresponding grid nodes, and those nodes formed a wireless sensor network to realize the long-term and unmanned observation of the environmental factors.

Given the cost of the current deployment of sensor networks is not low and the related theory and technologies are still in the research stage. Therefore, as shown in **Figure 2**, we first carried out the deployment of wireless sensor network on one grid node to analyze network performance in extreme and extraordinary conditions, explore availability of WSN demonstration for environmental monitoring and further feasibility of large-scale nodes deployment.

# 3. SYSTEM FRAMEWORK OF ECOLOGY-HYDROLOGY WIRELESS SENSOR NETWORK

In this paper, the system framework of Ecology-hydrology Wireless Sensor Network is divided into four layers that are Sensor Network Layer, Server Layer, Application layer and User Interface Layer.

Sensor Network Layer: Observing data packets collected by sensor nodes will be sent through the self-organizing multi-hop wireless network to distant Base Station and Gateway. The data packages could be automatically forwarded through the best path which was chose based on hop counts, node energy, RSSI, and other factors.

Server Layer: the data will be uploaded to the remote database server via transmission network such as satellite, GSM/CDMA/GPRS, and be analyzed, computed and visualized.

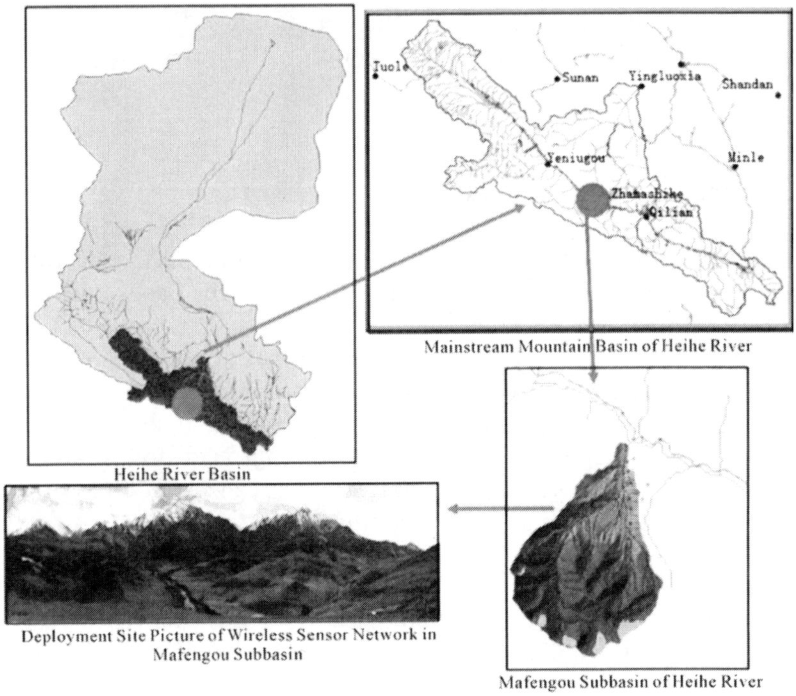


**Figure 1**. The location of Mafengou sub-basin in Heihe river basin.

Application Layer: a large number of continuous, intensive observational data can be collected by the deployment of Ecology-hydrology Wireless Sensor Network. Based on specific application scenarios, researchers could build a number of WSN applications.

User Interface Layer: users can easily access and download data through a browser. Administrators could accomplish the tasks such as node status monitoring, data collection and data visualization.

# 4. WIRELESS SENSOR NETWORK

As shown in **Figure 3**, according to the system framework of Ecology-hydrology Wireless Sensor Network, WSN is mainly constituted by nodes, base stations, transmission networks, gateways, servers and other components.

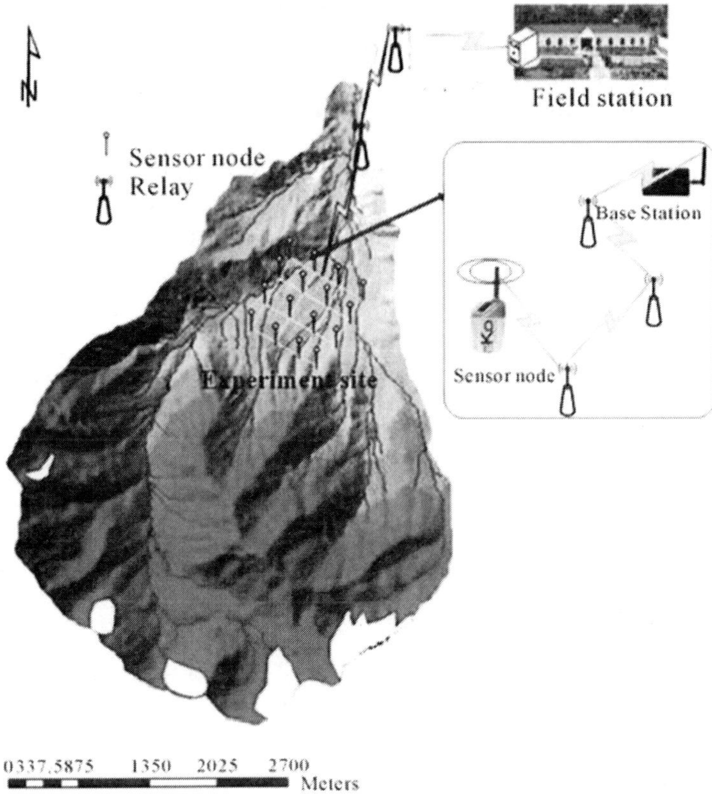


**Figure 2**. Deployment of the eco-hydrological wireless sensor network.

## 4.1. Nodes

We adopted a set of eKo nodes produced by Crossbow Technology Inc. which integrates an IRIS series processor or radio board and the 2.4 GHz dipole antenna. These nodes support the radio transmission of 2.4 GHz ISM band and could be serially connected up to 4 sensors. Its power is supplied by solar panels and three rechargeable Ni-MH batteries which could last three to four months uninterruptible power supply without sunshine. Data sampling frequency is set by the program into 15 minutes per time. As this data sampling frequency and without connecting sensor nodes, the average operating current is 0.4 mA. The sensor nodes and memory card in them will ensure proper operation at environmental conditions of −40°C to +60°C temperature and 0 RHI to +70 RHI humidity. It could work properly at temperature between −40°C and +70°C. Whenever reset the sensor nodes, they are required to be scanned and automatic identified. Taking into account the scenarios of site deployment, the node can transmit about 500 to 1000 meters of viewing distance by radio wave. In addition, the node also provides IP66-rated waterproof packaging to prevent dust and water droplets.

As shown in **Figure 4**, the nodes in the current field deployment serially connected soil temperature and humidity sensors of Watermark Company, environmental temperature and humidity sensors of Sensirion Company, and solar radiation sensors of Davis Company to monitor and collect information of relevant environmental factors.

In wireless sensor networks, nodes are generally far away from base stations, thus sensor's data which collected by the nodes could not be directly transferred to the base station and data forwarding was often relied on the relays. The hardware structure of relay was exactly the same as normal nodes but no sensors. In Hulugou sub-basin, the distance between the most remote node and the base station is 1.92 km in design. We adopted a self-developed routing protocol in this paper, and the transmission distance is about 700 m to 1000 m. In addition, considering equipment redundancy, we chose three relays which use a 7 dBi omni-directional antenna, transmitting in the 2.4 GHz ISM band, and adopt a joint supply of solar panels and Ni-MH battery.

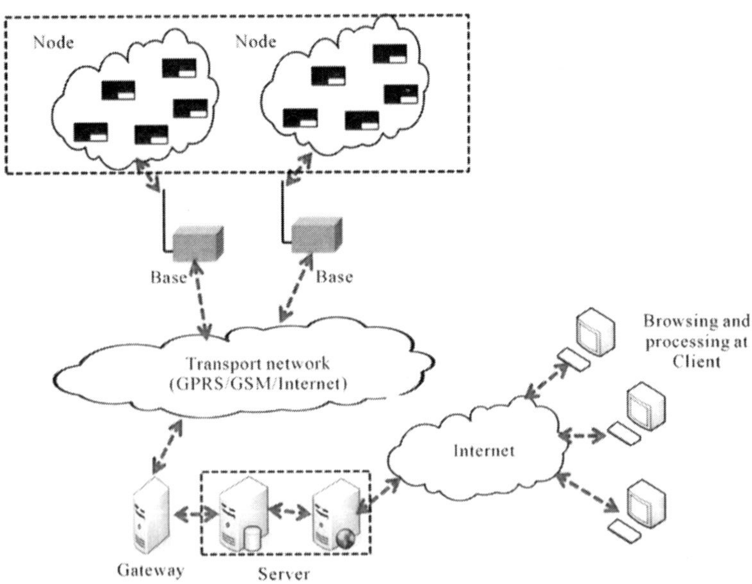


**Figure 3**. Wireless sensor network include sensor nodes, base stations, transmission networks, gateways, servers and other components.

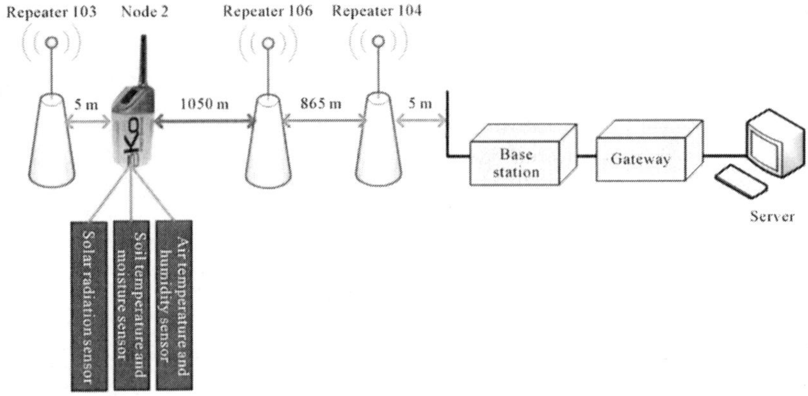


**Figure 4**. The network topography of watershed Eco-hydrological Wireless Sensor Network.

## 4.2. Base Station

The base station has a function of aggregating data collected by nodes and transmitting them to gateway node and the external network. It integrates a Crossbow IRIS series processor and a 2.4 GHz frequency radio band Omni-directional antenna. It also provides a USB access. The base station connects gateway equipment by an RJ45 cable for power. The average operating current of base station is about 30 mA and it could work in condition of 6°C to 40°C temperature and 10% to 80% humidity.

## 4.3. Transmission Network

After local computing and pre-processing of nodes, data packets collected by sensors were aggregated to the base station in the form of self-organizing multi-hop wireless network, then uploaded to the gateway by wired Ethernet, GPRS, GSM, Internet and so on. Through data analysis, data conversion, data processing, packet checksum calculation and information extraction, data in the gateway was saved to an SQLITE embedded database, then transferred to data center through the Internet at last.

## 4.4. Gateway

In this paper, the gateway acts as a small server. It has a built-in 266 MHz Intel IXP420 XScale processor, 8 MB program FLASH, 32 MB RAM and 2 GB flash disk. It could work properly at condition of 5 Voltage and 4 Watt power, 6°C to 40°C ambient temperatures and 10% to 80% humidity.

Built-in visualization Web application package in gateway could realize the remote data access and data management. Users can view real-time data, historical data and network-related configuration information through a Web browser at anytime and anywhere.

## 4.5. Servers

We set two tier servers for field observation station to save data from Sensor Network: One server locates in the field station that for saving data, pre-processing data and then submit the data to server deployed in data center. First server acts mainly as storage of data to extend the base capacity. The other server locates in the data center acts for further dealing with data, data analysis, data visualization, and sharing data through Web service.

## 4.6. Data Center and Communication Network

Our supercomputing environment adopts the high-performance parallel computer systems of Dawn 5000. It comprises four fat nodes and 64 blade servers which connected with InfiniBand high-speed network card and add up to 1152 cores AMD CPU, floating-point processing power is up to 10 trillion times a second. PC users could connect supercomputer environment and remote control the wireless sensor network of Eco-hydrology experiment field of Heihe river basin.

# 5. DATA ANALYSIS METHODS AND PERFORMANCE ANALYSIS OF SENSOR NODES

## 5.1. Data Problems and Data Quality Control Methods

### 5.1.1. Data Problems

If the data sensing components of a Wireless Sensor Network system, sensors, do not have a better accuracy, it will result in less accuracy data or error data. Data processing unit of sensor has a limited storage space and will affect the system transitional surge in the exchange of data processing capabilities to result in loss of metadata information. Lack of enough sensor nodes or the sensor nodes could not adequate cover the research area will lead to incomplete collection of sensor information. Excessive number of sensor systems will result in intensive competition between nodes and busy the data transmit channel between nodes and base station, leading to severe network congestion, noisy data and inaccuracy data. In addition, unsuitable data link layer protocols and data routing protocols of sensor network will bring loss and delay of a large number of sensor information packets. For the low temperature, low energy and other environmental factors, sensor network nodes deployed in harsh environments will make abnormal values in sensor data. These above problems in wireless sensor networks will reduce the quality of observing data.

### 5.1.2. Data Quality Control Methods

Data quality control process could take the following steps:

1) Find the problem data. Firstly check logs of data acquisition and sensor network, then look through trends and constraints in environmental factors and check whether the data was within normal range of variation. Finally, drawing maps and comparing the relationship between the observed data from traditional integrated environmental observing system of Hulugou sub-basin and from wireless sensor networks to find observation which does not meet the basic principles.

Secondly, review the trends of data. Temperature and humidity of soil, temperature and humidity of environment, solar radiation intensity in this paper are all the meteorological factors which change periodically, and error analysis using conventional methods could not identify the individual error data. For example, values of some elements may have steep changes in a short time, but its value is still between the maximum and minimum of the same day or month, thus it will result to difficult judging based on the error. Curve prediction model can be used to carry out anomaly trend of data.

Thirdly, review the constraints. Collecting these observed elements have their own constraints, such as maximum constraint, minimum constraint and so on. The data of observation elements over times should be greater than the minimum constraint and smaller than the maximum constraint.

2) Identify and classify causes of data problems. One type of problems is logic errors of node's collection program and unsuitable design of data link layer protocols, network routing protocols and other communication protocol stacks. Another type is hardware problems of sensor network caused by an uncorrectable error, and this type of problems could be reduced by replacement of equipment. The other one is the problems resulted by human factors, environmental factors and animal interference, the causes of these problems could be identified by setting up camera to monitor and analyze the regional image.

3) Correct and revise the problems. According to the above classification of problems, correct and revise the problems which could be corrected, backup and delete problems which could not be corrected.

## 5.2. Data Management and Sharing

A semi-structured distributed model of data management was adopted to manage data in this paper. As shown in **Figure 5**, the model is divided into two or more levels. The bottom layer is the sensor network which was composed by sensor nodes of limited capacity. Each node receives command from the agent layer, collects data and computes locally, and transfer data to the agent layer. The agent layer is composed of base stations and gateways. Agent layer receives data from sensor nodes, indexes the received data or receives requests from users and forwards control commands to sensor nodes to process distributed query processing and returns query results to the users.

## 5.3. Data Visualization Analysis

As shown in **Figure 6**, web services program in gateway supports visualizing display elements by pie chart, trend line and other visualization forms. Real-time network topology, map window, servers' information, nodes' information, and routing information visualization are provide in this web service, and the sensor node list shows all the sensor nodes to be managed in current local area network. The data obtained through the WSN are transmitted to the research station by immediate and automatic means, and then transported to the super computing system at our institute through optical fiber channel, and finally displayed as 2D, 3D images on the Web page after processing.

# 6. CONCLUSION AND DISCUSSION

## 6.1. Conclusion

The deployment and research of wireless sensor network in Hulugou sub-basin completed the following works: Establishing a set of wireless sensor network which composed by sensors, nodes, base stations, relays, network management system, servers for observing hydrological and ecological elements. Obtaining data related to soil moisture, soil temperature, air temperature radiation data of twelve months about Hulugou, and forming a technical system to get a longtime intensive observation data of eco-hydrological elements. To providing validation of new observation methods of sensor network of eco-hydrological process studies and environmental monitoring in extreme cold and high altitude environments to form a demonstration role. These application and research promotes eco-hydrological observations of Heihe River Basin extending from point observation to intensive observation extension and provides high-precision spatial data for constructing eco-hydrological model and preparing driving data for model simulation.

## 6.2. Discussion

Wireless Sensor Networks should be taken as an additional intensive data source of traditional fixed-point observation station to validate the data. High spatial and temporal resolution data of wireless sensor network could provide high-precision surface data for modeling remote sensing retrieval model and validation of retrieval data. In future work, the low-power node and sensor technology should be further consider and cooperate closely with experts in specific areas to enhance a comprehensive understanding of research objects and network deployment, and to design and deploy more suitable wireless sensor network for researches and methods for data collecting, analysis and presentation.

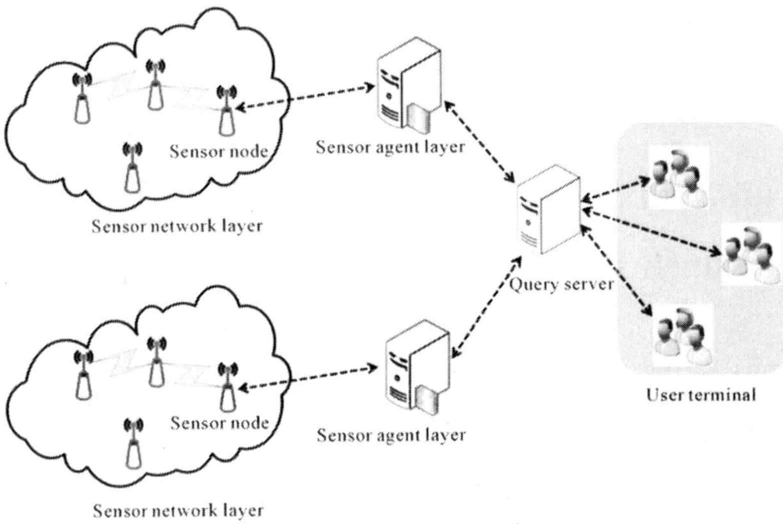


**Figure 5**. Semi-distributed data management model.

### 6.2.1. Problems

Although we made detailed designs and careful concerns to the problems of wireless sensor network deployment, but many challenges are still exist. Marcus Chang also summarized some lessons of monitoring in a high-Arctic environment [14]. We have gathered a lot of valuable experiences and lessons in addressing these challenges.

Long rainy weather will lead to less charge of a sensor node, and problem of packets loss will be more severe as energy depleting. The temperatures of Hulugou Subbasin are often below zero, it results in air of solar cells box condensing to water and shorting circuit of solar panels to bring a long-term packet loss. For the temperature impact, the clock of sensor nodes and relays will shift and will result in time synchronization inconsistencies for data transmission in a sensor network, the phenomenon of delays and packet loss will occur. In the development process of sensor network equipment and software, the principles of as simple as possible should be stressed, and code and algorithms must be able to match the specific application to avoid interaction between software, to reduce energy consumption a certain extent, and to reduce low storage capacity constraints of sensor nodes.

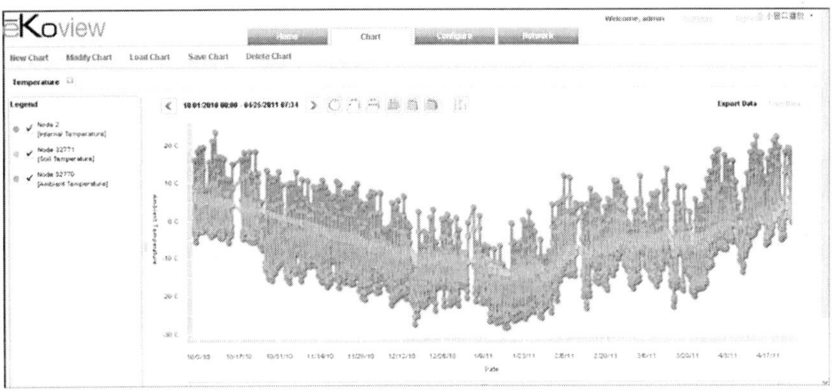


**Figure 6**. Trends of nodes internal temperature, environmental temperature and soil temperature overtime.

## 6.2.2. Experiences

Taking a serious investigation of environmental conditions of the monitored field before deployment, such as topography, climate conditions, the network signal coverage and so on, to study how environmental conditions affects the entire sensor network system. But we could not build a suitable model to fully express how environmental conditions affect the hardware and the entire system. Note that devices related the wireless sensor network should be labeled to inform the people our ongoing experiments.

Sensors of soil temperature and humidity should be saturated for all one night before deployment. If not, sensors buried in soil will lead to unstable and ineffective value for several days. The values would be normal until the probe of sensors were full saturated and mixed together with the surrounding soil.

For the lax of relay solar box packages, condensation water leaded to bring into the short circuit phenomenon. In addition, the light of relays and sensor nodes should not lit always, because the light could be a source of relatively large energy-consuming. Thus they should be shut off to save energy after the sensor finishing the role as a debugging tool. Many problems were encountered in carrying out observations in Wireless Sensor Network, and the experiences of addressing them will provide some reference for future work.

### 6.2.3. Main Concerns

As the features of limited energy, limited storage capacity and limited computing power of nodes in Wireless Sensor Network, low-power operating system and routing protocols should need to be developed. Routing protocols should be designed for some specific applications of sensor network, and there is no universal routing protocols suitable for any application. In addition, power supply for gateway should be fully protected to avoid data loss. Animals will also bring a bad effect to sensor network deployment, thus nodes bracket and cable should be reinforced.

### 6.2.4. Future Work

In the future, we will deploy a large number of sensor nodes in the long-term fine-grained Eco-hydrological elements observing environment to form lots of homogeneous or heterogeneous Wireless Sensor Networks. Large amounts of data produced by these networks will result in study need for a distributed sensor data access, integration, intensive calculations, and integration with the fixed-point observation data, simplify the operation of wireless sensor networks, visualize and effectively display the observed data through Web.

## ACKNOWLEDGEMENTS

This research was supported by the Special Project of Information of the Chinese Academy of Sciences, "EScience environment of Ecology-Hydrological model integrated research in the Heihe river basin (INFO-115- D01)", and the Incubation Foundation for Special Disciplines of the National Science Foundation of China (NSFC) (grant number: J0930003/J0109), and the "Ice, Snow and Permafrost environment Database (INFO-115- C01-SDB2-03).

## REFERENCES

1. L. H. Wang, et al., "Advance in Basin Eco-Hydrological Process Modeling," Advances in Earth Science, Vol. 24, No. 8, 2009, pp. 891-898.

2. P. Gong, "Progress in Recent Environmental Application of Wireless Sensor Network," Journal of Remote Sensing, Vol. 14, No. 2, 2010, pp. 387-395.

3. A. Zerger, R. A. Viscarra Rossel, D. L. Swain, T. Wark, R. N. Handcock, V. A. J. Doerr, G. J. Bishop-Hurley, E. D. Doerr, P. G. Gibbons and C. Lobsey. "Environmental Sensor Networks for Vegetation, Animal and Soil Sciences," International Journal of Applied Earth Observation and Geoinformation, Vol. 2, No. 5, 2010, pp. 303- 316.

4. W. R. Philip, A. G. Eric, F. A. Michael, C. F. Jason and C. H. Thomas, "Environmental Sensor Networks in Ecological Research," New Phytologist, Vol. 182, No. 3, 2009, pp. 589-607.

5. S. S. O. Burgess, M. L. Kranz, N. E. Turner, R. CardellOliver and T. E. Dawson, "Harnessing Wireless Sensor Technologies to Advance Forest Ecology and Agricultural Research," Agricultural and Forest Meteorology, Vol. 150, No. 1, 2010, pp. 30-37.

6. J. Yang, C. Zhang, X. Li, Y. Huang, S. Fu, et al., "Integration of Wireless Sensor Networks in Environmental Monitoring Cyber Infrastructure," Wireless Networks, Vol. 16, No. 4, 2010, pp. 1091-1108.

7. J. V Borre, D. Paelinckx, C. A. Mücher, L. Kooistra, B. Haest, G. D. Blust and A. M. Schmidt, "Integrating Remote Sensing in Natura 2000 Habitat Monitoring: Prospects on the Way Forward," Journal for Nature Conservation, Vol. 19, No. 2, 2011, pp. 116-125.

8. N. W. T. Quinn, R. Ortega, P. J. A. Rahilly and C. W. Royer, "Use of Environmental Sensors and Sensor Networks to Develop Water and Salinity Budgets for Seasonal Wetland Real-Time Water Quality Management," Environmental Modelling & Software, Vol. 25, No. 9, 2010, pp. 1045-1058.

9. X. Zhu, D. Li, D. He, J. Wang, D. Ma and F. Li, "A Remote Wireless System for Water Quality Online Monitoring in Intensive Fish Culture," Computers and Electronics in Agriculture, Vol. 71, No. 1, 2010, pp. 3-9.

10. S. Bonnot, S. Gianfranceschi, S. Smolders, M. Gilles, T. Holzer-Popp and M. Erlich, "GENESIS, a Flexible Solution for Building Networks of Collaborative Environmental Services," IFIP Advances in Information and Communication Technology, Vol. 359, 2011, pp. 195- 207.

11. C. Youn, S. Chandra, E. H. Fegraus, K. Lin and C. Baru, "TEAM Network: Building Web-Based Data Access and Analysis Environments for Ecosystem Services," Procedia Computer Science, Vol. 4, 2011, pp. 146-155.

12. Y. Cheng, D. Y. Eun, Z. Qin, M. Song and K. Xing, "The Design of a Wireless Sensor Network for SeismicObservation-Environment Surveillance," Lecture Notes in Computer Science, Vol. 6843, 2011, pp. 157-167.

13. R. Chen, et al., "Hydrology, Ecology and Climate Significance and Its Research Progress of the Alpine Cold Desert," Advance in Earth Science, Vol. 25, No. 3, 2010, pp. 255-263.

14. M. Chang and P. Bonnet, "Monitoring in a High-Arctic Environment: Some Lessons from MANA," IEEE Pervasive Computing, Vol. 9, No. 4, 2010, pp. 1-10.

8

# WIRELESS SENSOR NETWORK COMPONENTS FOR AIR POLLUTION MONITORING IN THE URBAN ENVIRONMENT: CRITERIA AND ANALYSIS FOR THEIR SELECTION

**Vendula Hejlová, Vít Voženílek**

Department of Geoinformatics, Palacký University in Olomouc, Olomouc, Czech Republic

## ABSTRACT

Wireless sensor networks (WSNs) are fast evolving technology for collecting data in real time. Every wireless sensor network (WSN) is consisted of technical and software components which have to refer to the selected application. The paper focuses on the selection of WSN components. The WSN will be situated in the center of Olomouc City (OWSN). It will focus on measurements of harmful air pollutants and selected basic meteorological elements. The criteria for selection of WSN components including the most important parameters will be chosen and the final evaluation of the option utility will be made on the basis of multicriteria decision making process.

## KEYWORDS

Components of Wireless Sensor Network; Criterion; Multicriteria Analysis; Sensor Node

# 1. INTRODUCTION

WSNs are new fast evolving technology that has been successfully applied to the monitoring of selected phenomena in the wide range of fields trespassing to the socioeconomic and environmental domain. The WSNs in socioeconomic area are used for monitoring health state, park places, cars, energy in buildings and logistics. In the environmental direction they are used for detection of landslides, forest fires, air pollutants, volcanic activity, water quality, biodiversity, microclimate or irrigation planing [1-3].

The authors designed the WSN which will be situated in the centre of Olomouc city. This WSN will be used for monitoring air pollutants and selecting basic meteorological elements which influence concentration and dispersion of air pollutants.

The aim of this paper is to determine criteria for node (motherboard), sensor boards/sensors and gateway selection of the designed WSN. Technical components, their parts and software are evaluated on the basis of criteria which are defined for nodes (with included demands on the sensor boards/sensors) and gateway. Demands on software were included in the criteria for the selection of hardware components.

Every criterion has its own importance. The criteria for nodes are divided into eight groups which are rated according to their importance. The criteria for selection of gateway are divided into seven groups. Every group of criteria for nodes and gateways includes particular criteria. The ideal parameter for every criterion of Olomouc WSN (OWSN) is suggested. The ideal parameters of chosen criteria should be fulfilled in the case of the ideal option.

Manufacturers of WSN components and their products were searched. Every searched product (option) had to have fulfilled parameters related to the chosen criteria. Due to this step all options with their parameters were comparable because they could be evaluated on the basis of the same criteria with fulfilled parameters. Every criterion was evaluated by its importance in the selecting process. Finally, the weight computation and subsequent analysis were provided and the most suitable option was selected.

## 2. WIRELESS SENSOR NETWORK COMPONENTS

Every WSN consists of two major groups of components—hardware (technical platform) and software.

The technical platform of the WSN is defined by nodes (motherboard) and gateway (base station). Every node consists of several parts: microprocessor, battery, communication unit, memory, built-in sensors and interface for connecting sensor boards/sensors [4]. **Figure 1** shows the technical parts of the node. The sensors boards/sensors can be connected to the node [5]. Some nodes gather data from the other nodes and send them to the gateway. These nodes are called routers. The electromagnetic radio waves are used for the communication among nodes and between nodes and gateway. In recent research a vast number of WSN communication protocols operating in radio frequency has been developed [3]. Data are stored in the gateway or sent to the server for the long time storage. The nodes placed into the outdoor conditions have to have an ingress protection which protects the hardware components against the external influences [5].

Software of nodes controls hardware component, manages system resources and allows the use of applications [6].

Teams of experts deal with the application of the WSNs all around the world. WSNs for monitoring of harmful air pollutants have been tested and applied in Spain, Great Britain and Serbia. The new WSN is planned to be built in the Netherlands, Assen City. This technology has been mostly applied (except for Europe) in the USA, Taiwan, China and India.

The WSN for monitoring of air quality in Salamaca, Spain, is good example of effective network. It was built in the cooperation with CARTIF, European BIC Network, Salamanca City Hall and Research Advisory Group P&G. This WSN is built on the components produced by the Spanish Company Libelium. Every node includes sensors for measuring temperature, relative humidity, $NO_2$, $O_3$, CO, PM and noise level. Data are sent to the gateway from which they are resent to the server for long time storage. Nodes can react to overstepping of the allowed limits which are set for every measured element. If the limit is exceeded, the emergency message is sent to the gateway and the warning is issued. This WSN was created in 2011 within the RESCATAME project using capabilities of the Spanish producer [7,8].

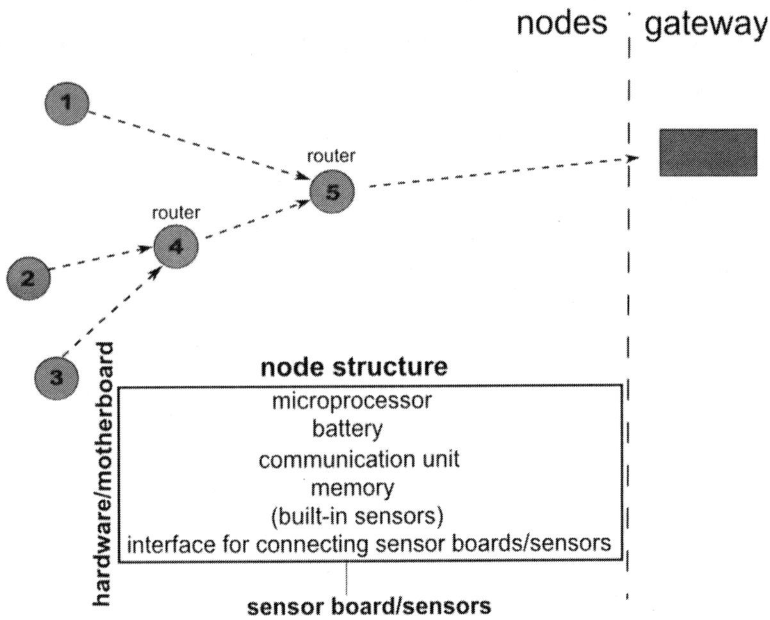


**Figure 1**. Technical parts of the wireless sensor node.

The specially developed nodes for monitoring of air pollution were used in London in 2007. The sensors for monitoring $SO_2$, NO, $NO_2$, $O_3$, $NH_3$ and benzene were connected to the nodes [9]. The prototype of the WSN which used GUSTO nodes was tested in the Cambridge University Campus in 2008. The nodes carried the sensors for monitoring CO, $CO_2$, $NO_2$ and GPS receiver. Data were sent from the nodes to the central storage through the mobile network [10, 11].

In Serbia there are also used nodes with sensors produced by Libelium. The nodes are situated on the moving buses so that the movement of the bus can be traced. The information about position of buses can be displayed in a mobile application [12].

The great emphasis is placed on the usage of WSNs for monitoring of air pollutants in Taiwan. The nodes, called OCTOPUS II, were developed in cooperation between government and universities. These nodes were equipped with sensors produced by e2v for CO measureing [11].

The high attention for using WSNs has been focused in Indian and Chinese towns. The increasing number of cars causes more harmful pollutants in the air. In Indian Hyderabad City there are used the nodes and sensors produced by Libelium [13]. The motherboards Mica2 which are produced by Crossbow Company are used for monitoring in Nagpur City, India. The sensors were specially developed for this application [14]. The University in Cinncinati is oriented on the monitoring of air pollutants using the MICAz nodes which are produced by Crossbow Company. They use the sensors of KWJ Engineering [15].

In above-mentioned studies there was given only one reason for the selection of components. This reason pointed to the type of application of the WSN and the compatibility of sensors measuring air pollutants and the motherboard. The greatest attention in the process of wireless sensor components selection is aimed at nodes. If the commercially offered options did not meet requirements for the application, the nodes were developed as a part of project as in the case of London wireless sensor network [10,11].

## 3. THE SELECTION OF WSN COMPONENTS

The selection of the WSN components is very important because it is a crucial decision. On its basis the measurements of harmful air pollutants in the urban environment will be performed. Pre-selection steps involve:

- determination of the purpose of building
- determination of the number of nodes
- choosing the localization of the nodes
- searching for the producers (analyzing commercially offered components and comparing their functionality)
- defining the criteria for the evaluation of selected options
- defining the ideal option
- comparing criteria parameters of offered options to the ideal (suggested) one (intended configuration)

# 4. CRITERIA FOR THE WIRELESS SENSOR NETWORK SELECTION

Criteria were formulated according to the aim and purpose of the Olomouc WSN (OWSN). Suitably selected criteria serve as a basis for the determination of the fitted option with using the multicriteria analysis decision making process. This method is used for the selection of the most suitable product (option) from the possible solutions [16, 17].

A criterion in a decision making process is called the scale and it is used for the comparison of the chosen option parameters. Every option should have fulfilled the criteria parameters. Criteria have to cover the most suitable parameters which have to be completed so that hard expressible criteria can be selected. The importance of every criterion which is a basis for the determination of its weight have to be indetified. The weight shows the importance of the selected criteria in the decision making process and it is written as a positive real number [16,18]. There is always an uncertainty in the process of determination of criteria [19].

The weights of criteria were determined by two methods: The Order and The Compensation Method. Weights of every criterion determined by these methods were compared. The method which was selected for the determination of the final result is called The Weighted Sum Method. The selected method is ordered to the group of methods based on the weighted average of evaluation. The final result was computed with The Aggregated Utility Function which is one type of the advanced computational algorithms [16,18]. The weights computed by The Order and The Compensation Methods were used to the computation which determines the order of options.

Factors like node size, price and energy consumption play the most important role in the selection of the WSN nodes. According to [20] "the nodes could be in some cases smaller than 1 $cm^3$, with the weight lower than 100 g, cheaper than 1 USD and the energy consumption lower than 100 μW." Sensor boards/sensors have to complete the requirements for the wireless sensor application and to measure elements which are suitable for the chosen task. The gateway has to be flexible, with a large memory size and it has to have a large communication range with the possibility of connecting through ethernet to the server.

The computation of the best option was made for nodes. The requirements for software and the sensor boards/sensors were added to the ones which are related to the nodes. It is presumed that the other

WSN components will be provided by the same company like the selected nodes. If the result will not be unambiguously the analysis of the other components and companies will be provided. The most suitable commercially offered option will be applied for the monitoring in Olomouc City.

## 4.1. Criteria for Nodes Selection

The nodes are the basic building unit of the WSN so that the majority of requirements are related to this kind of hardware component. Eight groups of criteria for nodes were defined and they can be seen in **Table 1**. These groups were thematically divided. The importance of each criterion was specified on the basis of defined parameters for the ideal option and the following consultation with experts.

The first group of criteria deals with the offered sensor boards/sensors because the sensors for the measurement of harmful air pollutants have to be connected to the motherboard. The first criterion of this group relates to the possibilities of connecting sensors which measure elements fulfilling the purpose of the application. It is evaluated if the sensors are produced by the same producer, like the motherboard, or if they have to be bought from another producer. The count of sensors connected to the motherboard shows the maximum amount of observed elements. Sensor measurement interval differs from elements because some of them change more quickly than the others. Calibration of sensors is very important because it is made in the laboratory conditions with the knowledge of calibration curves so that it should be provided by the producer of sensors. Sensors can be built in the motherboard. These sensors are not suitable for external monitoring because they are covered in the enclosure and they measure inside nodes parameters.

The second group of criteria is called communication models. The first and the most important criterion from this group is that nodes have to communicate not only with the gateway (or server) but with one another. Nodes sending data directly to the gateway or sever do not fulfill all requirements for the WSN. Their configuration can be considered to the star topology with the long distance data transmission. In the most of cases they communicate through mobile or internet network. If the protocols with the shorter range are used, then all nodes have to be placed in a short distance from the gateway [21]. The communication among the nodes can be performed on short or long distances depending on the used protocol. The nodes which send data through the other nodes are called routers. The communication between nodes and gateway uses the same protocol as the communication among nodes. All measured data are gathered in the gateway and sent to the server. This type of communication can be performed for short or long distances. The most common and the safest communication between gateway and server is performed using ethernet. The selected type of the communication deals with the communication range which shows the maximal range for data sending. Topology is the last criterion in this group. Two most commonly used types of topology are tree and mesh topology.

Energy sources are included in the third group of criteria. The most important criterion in this group deals with the possibilities of energy sources. There are a lot of ways how to charge the nodes. The first type of charging is by the various types of batteries, the second one by electricity network and the third one by USB or some other computer compatible connector. The power options are limited in the environmental conditions. The most commonly used form of charging includes batteries. The battery life depends on the energy consumption which depends on the selected mode and its occupancy. The major influence on the battery life has the fact if the node is the router or not. It is important to know if there is a possibility of charging battery by solar panel. Due to the solar panel battery of the nodes is continuously charged.

The fourth group of the criteria is oriented on the availability of nodes to resist the environmental influences such as rain, dust, temperature extremes etc. The storage temperature and the protection of nodes belong to these demands. It is necessary to obtain data during the whole year in all kinds of weather conditions. The external influences depend on the part of the world where the nodes are installed. The protection from the effects of the outer environment is made by the enclosure which must not have an influence on the measured values but it has to protect the hardware of the node.

The fifth group is based on the technical aspects of the WSN. The first criterion in this group is called the level of programming skills. This level can be low, middle or high. It depends on the pre-installed software parts in the node from the producer. The user has to have the option to interfere to the program code and suit it to its own application. The criterion on memory size is very important when the data are stored in the node for a longer time and they are not directly sent to the gateway. Input and output capabilities relate to the possibilities of accessing to the node.

**Table 1**. Criteria for nodes selection.

| Criteria Group Number | Criteria Group | Criteria for Nodes Selection |
|---|---|---|
| 1 | Sensor boards/sensors | sensor board for monitoring of air pollutants from the same producer |
| | | possibilities for connecting sensors for monitoring of pollutants |
| | | count of connected sensor boards/sensors |
| | | sensor measurement interval [min] |
| | | calibration |
| | | built-in sensors |
| 2 | Communication models | communication among nodes |
| | | communication possibilities |
| | | communication range [m] |
| | | topology |
| 3 | Energy sources | energy source |
| | | battery life [year] |
| | | energy consumption |
| 4 | Environmental resistance | storage temperature [°C] |
| | | protection |
| 5 | Technical aspects | programming skills |
| | | software |
| | | memory size [KB] |
| | | I/O |
| 6 | Expansion options | flexibility |
| | | sensor boards/sensors change |
| | | GPS |
| 7 | Physical parameters | size [mm] |
| | | weight [g] |
| 8 | Price | price for one node [€] |

The possibilities of WSN extension are summarized in the sixth group. Criteria like flexibility, possibility of sensor changing and adding a GPS unit are summarized in this group. The WSN can be extended by the number of nodes in the same place or it can expand to the larger area. Sensors or GPS unit can be added to the node.

Physical parameters are summarized in the seventh group of criteria. The first criterion is connected to the size of the node. The second one focuses on the weight of the node. These two parameters are important in the case that the nodes have to be hidden so that they do not arouse the attention of vandals [22].

The eight group of criteria is aimed at the price of node. The price of one node depends on a lot of demands such as application because this demand depends on the number of nodes in the field and the selection of sensors for the measured values of the elements. The price of one node ranges from 99 to 600 EUR. The price depends on the equipment of node and its measurement capabilities.

## 4.2. Criteria for Gateway Selection

Requirements for the gateway selection are very similar to those which are placed on nodes (**Table 2**). Demand on sensors selection does not have to be solved in the case of gateway. Criteria for gateway selection are divided into seven groups.

The first group of criteria is aimed at the communication models. The possibilities of communication between nodes and gateway, gateway and server are involved in this group. Topology possibilities are taken into account.

Energy sources are included in the second group of criteria. The first criterion deals with power options. Gateway can be powered by electricity, because it can be situated in the indoor environment, because this component does not measure outside parameters. Gateway can be also powered by batteries and in this case it is necessary to consider battery life which should be high with low energy consumption.

**Table 2**. Criteria for gateway selection.

| Criteria Group Number | Criteria Group | Criteria for Nodes Selection | Increasing/Decreasing/Boolean Preference |
|---|---|---|---|
| 1 | Communication models | | |
| | | communication possibilities | max |
| | | communication range [m] | max |
| | | topology | type |
| 2 | Energy sources | | |
| | | energy source | max |
| | | battery life [year] | max |
| | | energy consumption | min |
| 3 | Environmental resistance | | |
| | | storage temperature [°C] | max |
| | | protection | yes |
| 4 | Technical aspects | | |
| | | programming skills | min |
| | | software | type |
| | | memory size [KB] | max |
| | | security | max |
| 5 | Expansion options | | |
| | | flexibility | max |
| | | sensor boards/sensors change | yes |
| | | GPS | yes |
| 6 | Physical parameters | | |
| | | size [mm] | min |
| | | weight [g] | min |
| 7 | Price | | |
| | | price for one gateway [€] | min |

The third group of criteria deals with resistance of gateway against the external influences. The criteria dealing with operational temperature and protected encapsulation are sorted into this group. The operational temperature must correspond with conditions of the place where the gateway is situated. Only a small number of gateways is suitable to be placed into the outdoor conditions. Encapsulation is necessary if the gateway is located in the external conditions.

Technical aspects are involved in the fourth group of criteria. The first criterion in this group deals with the programming skills of the user. Basic interface of gateway has to be programmed. There have to be possibility of intervention to the program code. The memory size is considered in the case of gateway. Measured values can be stored in gateway for a long time, before they are sent to the server. The security has to be taken into account because in the gateway are stored a lot of data and they must not be lost.

The fifth group of criteria is aimed at the possibilities of expansion. Flexibility and extension by additional components like GPS receiver are involved in this group.

Physical parameters like size and weight of a node are included in the sixth group of criteria. The small size and weight are important if the nodes are situated in the external conditions and the gateway must not be immediately spotted or have to be placed in the higher position like roof or the top of traffic lamp.

The last group of criteria deals with price. The price depends on possibilities of communication which enables to connect the gateway with nodes and server and memory size.

# 5. REQUIREMENTS FOR THE WIRELESS SENSOR NETWORK COMPONENTS IN OLOMOUC CITY

The major aim of Olomouc WSN (OWSN) is the monitoring of air pollutants. The ideal parameters for every criterion will be suggested only for the nodes which will be further evaluated. This WSN has to complete the criteria evaluated by the highest values. Every criterion has suggested the ideal parameter which is shown in **Table 3**. Fifteen nodes are planned to be placed in the centre of Olomouc City. These nodes will be situated around busy roads and in the nearby park. Their area of localization is shown in **Figure 2**.

The most significant requirements are included in the first four groups of criteria. The first group is the most important because it solves if the sensor boards/sensors measuring the air pollutants can be connected. The second group shows the possibilities of communication and the third one examines the energy possibilities. The possibility of encapsulation is very important for the external monitoring.

Harmful air pollutants and basic meteorological elements observed by the OWSN should be in the ideal case: $CO_2$, $NO_2$, $O_3$, CO, $CH_4$, temperature, relative humidity, solar radiation, wind speed and wind direction. The list of ideally observed elements was made in the cooperation with the Czech Hydrometeorological Institute, Ostrava Office. The distribution, concentration and dispersion of the chosen air pollutants and meteorological geographical elements in the centre of Olomouc City will be observed.

Another significant request belongs to the second group of criteria called communication models. In this criterion there has to be confirmed the requirement for the possibilities of communication non licensed area in the Czech Republic (3.4 or 5 GHz). The usage of Zigbee protocol is necessary. This protocol is used for short and middle distance transmissions. The distance between the nodes in the OWSN should be at least 300 meters.

In the third group of criteria which is about energy sources there is the most important requirement for using battery power as an energy source.

The fourth group deals with environmental resistance which is very important when the nodes are placed in the external conditions.

## 6. COMMERCIALLY OFFERED COMPONENTS

A lot of companies offer WSN components but only a few of them can offer nodes which communicate with each other. The combination of components from different companies could bring problems with the installation and subsequent commissioning due to incompatibility among nodes and sensor boards. All products offered by ten companies were taken into account (**Table 4**).

The most suitable options were selected from 28 types of nodes, 52 sensor boards/sensors and 9 gateways. All these components were evaluated on the basis of above selected criteria. The highest attention is aimed at the sensor nodes because they create the most important part of the whole system. The chosen nodes are completed with sensor boards/sensors from the same producer.

The chosen criteria were not fulfilled in the case of all products (options) because the needed information is not provided by the producer. Nodes which had fulfilled at least 80% from the chosen criteria were selected to the evaluation. 15 nodes have completed this requirement. They are produced by 5 companies.

The criteria were used for the evaluation of producers of WSN components. It is necessary to know how they behave to the potential customer. It was evaluated if the distributor communicates with the user. Scale from 1 to 5 was chosen to evaluate how many days the sending answer has lasted. If the answer was not accepted in five days this company was evaluated by number 5. The second criterion is aimed at the structure of the answer. The answer can be either blunt and only correspond to the desired questions, or may contain additional questions towards the user and additional information. The evaluation was made in the scale from 1 to 5 (**Table 5**).

# 7. RESULTS

The criteria for selecting the most suitable components were simplified. This step was made because the parameters of all options have to be easily comparable. The utility computing was made for nodes which had at least 80% of criteria fulfilled.

**Table 3**. Parameters suggested for the OWSN.

| Criteria Group Number | Criteria Group | Criteria for Nodes Selection | OWSN requirements |
|---|---|---|---|
| 1 | Sensor boards/sensor | | |
| | | sensor board for monitoring of pollutants from the same producer | yes |
| | | possibilities for connecting sensors for monitoring of pollutants | - |
| | | count of connected sensor boards/sensors | min 6 |
| | | sensor measurement interval [min] | 15 |
| | | calibration | yes |
| | | built-in sensors | thermometer, akcelerometer |
| 2 | Communication models | | |
| | | communication among nodes | yes |
| | | communication possibilities | Zigbee protocol, communication in the range 3.4 GHz or 5 GHz |
| | | communication range [m] | 300 |
| | | topology | tree |
| 3 | Energy sources | | |
| | | energy source | battery |
| | | battery life [year] | one in the operation |
| | | energy consumption | the lowest |
| 4 | Environmental resistance | | |

| | | | |
|---|---|---|---|
| | | storage temperature [°C] | −30 - +40 |
| | | protection | yes |
| **5** | Technical aspects | | |
| | | programming skills | low |
| | | software | open source |
| | | memory size [KB] | 600 |
| | | I/O | must have USB connector |
| **6** | Expansion options | | |
| | | flexibility | yes |
| | | sensor boards/sensors change | yes |
| | | GPS | no, but there must be possibility for buying |
| **7** | Physical parameters | | |
| | | size [mm] | 100 × 100 × 100 |
| | | weight [g] | 100 |
| **8** | Price | | |
| | | price for one node [€] | 300 |

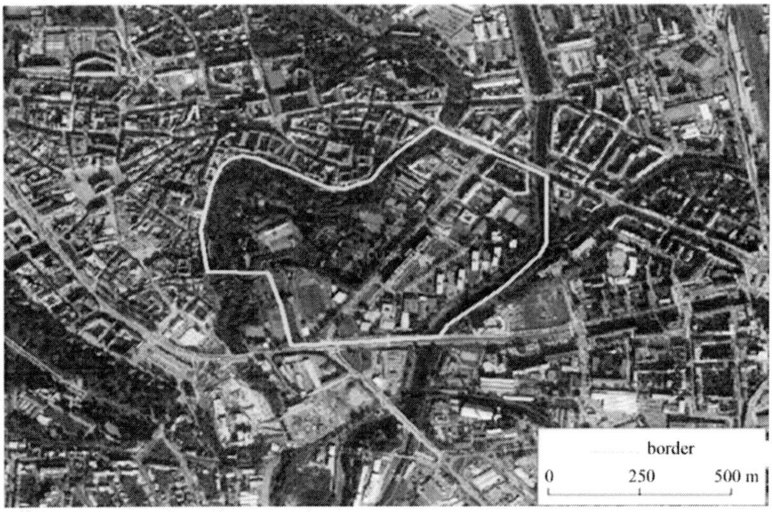


**Figure 2**. Area covered by nodes.

**Table 4**. Major producers of WSNs.

| Producer | State | Producer | State |
|---|---|---|---|
| **Libelium** | Spain | **Dust Inc.** | USA |
| **Memsic** | USA | **CSIRO** | Australia |
| **Crossbow** | USA | **Sun Microsystems** | USA |
| **TinyNode** | Switzerland | **Shimmer** | Ireland |
| **AdvanticSys** | Switzerland | **SilliconLabs** | USA |

The weights of the criteria which come from their importance were determined. Two kinds of methods were used for the determination of weights. The first method is known as The Rank Based Method, the second one as The Compensation Method. The weights determined by these methods can be compared because they were computed for all criteria. Their values are shown in **Table 6**and they are graphically compared in **Figure 3**. The highest values of weights were computed in the cases of the most important criteria. The differences between weight values computed by two methods are not so significant in the cases of middle important criteria. The most important criteria have higher weight values computed by The Rank Based Method. The least important criteria have higher values of weights computed by The Compensation Method. The highest difference between values computed by these two methods was in the case of criteria called Sensor boards/sensors from the same producer.

The Utility of the selected 15 options was computed by The Weighted Sum Method and The Aggregation Function. The utilities of the selected options with their producers are shown in **Table 7** and compared in **Figure 4**. The node called Waspmote Plug&Sense has the highest utility value of all options. This node is produced by the Spanish Company called Libelium. Nodes which have the second highest value of utility are produced by the same company as the first option and they are called Waspmote PRO nodes. The basic parameters of these two kinds of nodes are the same but the evaluation of Waspmote PRO was reduced by the criterion which includes node encapsulation. The third highest utility was reached by the nodes produced by the AdvanticSys Company. This company offers two kinds of nodes which reached the same result. The reason for the lower utility is the fact that this node does not have the encapsulation as the first mentioned option. Only these two companies offer nodes and sensor boards/sensors which measure the amount of air pollution. The requirement for the sensor boards/sensors measuring the air pollution from the same producer as the node has a high weight and that is the reason why these products are on the top of the result list. The eKo nodes are produced by Memsic. These nodes are encapsulated so they can be directly placed in the outdoor conditions. This criterion was in the overall evaluation highlighted. The node called MICA2 has the lowest utility. This node communicates in the spectrum which is not allowed in the Czech Republic. Requirement for the communication in the allowed spectrum has a high weight value. MEP-SYS are nodes which have the second lowest utility. These nodes are encapsulated but it is not possible to change their sensors and the sensors for the monitoring of harmful air pollution.

**Table 5**. The evaluation of the producers with at least 80% of criteria fulfilled.

| | | Libelium | Memsic | Crossbow | AdvanticSys | TinyNode |
|---|---|---|---|---|---|---|
| 1 | Communicate company with user | 1 | 1 | 5 | 4 | 5 |
| 2 | Structure of the answer | 1 | 1 | 5 | 2 | 5 |

**Table 6**. The weights of criteria determined by two methods.

| Criteria Group Number | Criteria Group | Criteria for Nodes Selection | Increasing/Decreasing/ Boolean Preference | Weight Rank Based Method | Weight Comp. Method |
|---|---|---|---|---|---|
| 1 | Sensor boards/sensors | | | | |
| | | sensor board for monitoring of pollutants from the producer | yes | 0.09 | 0.070 |
| | | possibilities for connecting sensors for monitoring of pollutants | - | 0.076 | 0.066 |
| | | calibration | yes | 0.047 | 0.055 |
| | | built-in sensors | yes | 0.023 | 0.039 |
| 2 | Communication models | | | | |
| | | communication among nodes | yes | 0.090 | 0.070 |
| | | Zigbee protocol | yes | 0.076 | 0.064 |
| | | communication in 3.4 GHz or 5 GHz | yes | 0.083 | 0.069 |
| | | middle values of communication range [m] | yes | 0.073 | 0.063 |
| 3 | Energy sources | | | | |
| | | battery powered | yes | 0.063 | 0.059 |
| | | battery life [months] | max | 0.060 | 0.057 |
| | | is solar panel possible | yes | 0.063 | 0.058 |
| 4 | Environm. resistance | | | | |

| | | | | | |
|---|---|---|---|---|---|
| | | is in operational range interval −30 - +40 [°C] | yes | 0.050 | 0.056 |
| | | protection | yes | 0.050 | 0.055 |
| **5** | Technical aspects | | | | |
| | | programming skills | min | 0.040 | 0.048 |
| | | memory size | max | 0.030 | 0.049 |
| **6** | Expansion options | | | | |
| | | GPS, change of sensors, flexibility | yes | 0.030 | 0.042 |
| **7** | Physical parameters | | | | |
| | | size [mm] | min | 0.020 | 0.020 |
| | | weight [g] | min | 0.013 | 0.021 |
| **8** | Price | | | | |
| | | price [€] | min | 0.023 | 0.038 |

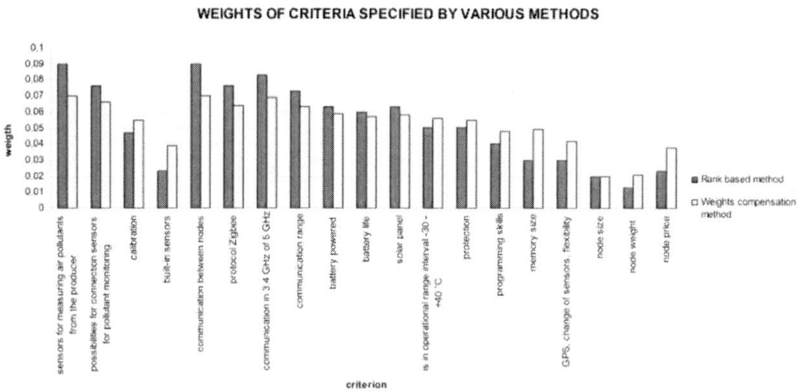


**Figure 3**. Weights of criteria determined by different methods.

# 8. CONCLUSION AND DISCUSSION

The WSN components selection for the particular application is a very important part of the whole project and certain attention must be paid to it.

More methods for the weight determination were applied so that the results could be compared. It is difficult to decide which method gives better results because in the higher and lower values there are significant differences in the weight values. The values of weights are similar in the middle part.

Multicriteria decision making process offers methods that are suitable for selection of the best option. This option meets the largest spectrum of the chosen criteria. The utility of the options was determined by The Weighted Sum Method including Aggregated Utility Function. The difference between the best and the worst option using The Order Weight Method is 0.41. This is a big difference and it shows that the worst option is not suitable for the requested monitoring. The best option utility value is 0.75 and this option fulfills the criteria evaluated by the highest values. The Waspmote Plug & Sense nodes represent the best option for this particular application. Not only nodes but the other WSN components compatible with these nodes are highly evaluated. Waspmote Plug&Sense node complies all the weight most evaluated criteria such as gas sensors from the same producer and communication using Zigbee protocol. These nodes are directly applicable in the exterior conditions. The disadvantages of these nodes are the size and the weight (**Figure 5**).

**Table 7**. The utility of selected options.

| Nodes name | Producer | Utility (The Rank Based Method) | Utility (The Compensation Method) |
|---|---|---|---|
| Waspmote PRO | Libelium | 0.710 | 0.712 |
| Waspmote Plug & Sense | Libelium | 0.749 | 0.763 |
| MICAz | Memsic | 0.468 | 0.460 |
| LOTUS | Memsic | 0.572 | 0.578 |
| TelosB | Memsic | 0.544 | 0.562 |
| IRIS | Memsic | 0.486 | 0.488 |
| IRIS (eKo Standard) | Memsic | 0.572 | 0.560 |
| IRIS (eKo Long Range) | Memsic | 0.599 | 0.580 |
| IMOTE2 | Crossbow | 0.466 | 0.453 |
| MICA2 | Crossbow | 0.336 | 0.358 |
| MEP-SYS multisensor | Crossbow | 0.354 | 0.436 |
| MEP-SYS temp-hum | Crossbow | 0.355 | 0.437 |
| TinyNode586 | TinyNode | 0.558 | 0.568 |
| XM100 | AdvaticSys | 0.638 | 0.638 |
| CM5000 | AdvaticSys | 0.637 | 0.637 |

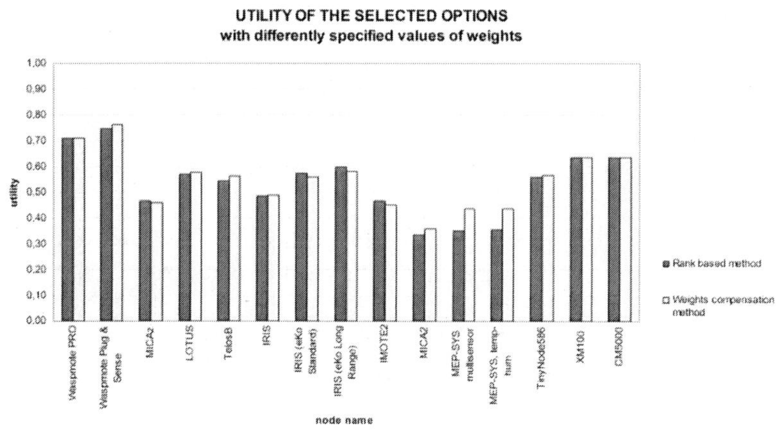


**Figure 4**. Utility of selected options.

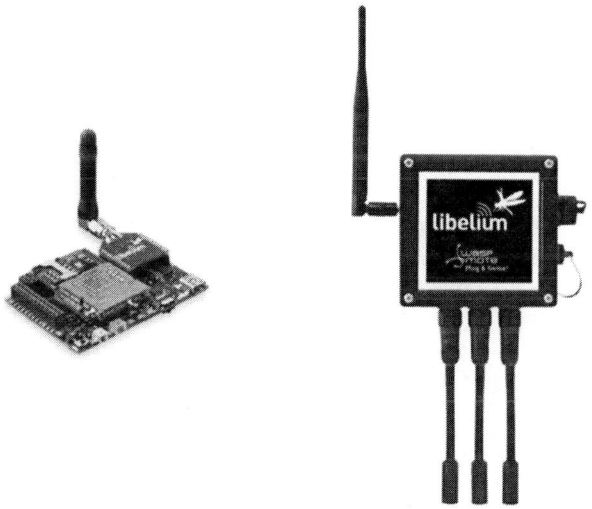


**Figure 5**. Waspmote Plug & Sense.


# ACKNOWLEDGEMENTS

The autohors acknowledge the support of the European Social Fund and the state budget of the Czech Republic (Project No. CZ.1.07/2.3.00/20.017). The authors also acknowledge the support of the Internal Grant Agency of Palacký University in Olomouc (Project No. PrF 2013 024).

## REFERENCES

1. A. Asín and D. Gascón, "50 Sensor Application for a Smarter World," 2013. http://www. libelium. com /top_50_iot_ sensor_applications _ranking/pdf

2. F. L. Lewis, "Wireless Sensor Networks," 2004. http://arri.uta. edu/acs/networks /WirelessSensorNetChap04.pdf

3. C. Murthy and B. S. Manoj, "Ad Hoc Wireless Networks: Architectures and Protocols," Prentice-Hall, Upper Saddle River, p. 857.

4. S. Büttrich, "Wireless Sensor Network," 2010. https://blog.itu.dk/SPVCE2010 /files/2010/11/ introduction_to_ wsn_-_lecture_slides_spvc2010_l.pdf

5. A. Grilo, "Wireless Sensor Networks," Single Node Architecture. 2007. http://comp.ist.utl.pt/ece-wsn/doc/slides/sensys-ch2-single-node.pdf

6. M. Podhorský, "Bezdrátové Senzorové Sítě v Průmyslové Praxi," Bachelor Thesis. VUT Brno, Brno, 2012. https://dspace.vutbr. cz/handle/ 11012/7764?show=full

7. A. Asín, "Sensor Network to Monitor Air Pollution in the Cities," 2010. http://www.libelium.com/smart_cities_wsn_air_pollution/

8. A. Bielsa, "Smart City Project in Salamanca to Monitor Air Duality and Urban Traffic with Waspmote," Libelium, Spain, 2011. http://www.libelium.com/Smart_City_Air_Quality_Urban_Traffic_W aspmote

9. M. Yajie, M. Richards, M. Ghanem, Y. Guo and J. Hassard, "Air Pollution Monitoring and Mining Based on Sensor Grid in London," Sensors, Vol. 8, No. 6, 2008, pp. 3602-3621.

10. K. Leung, "Mobile Environmental Sensing System," 1998. http://www. commsp. ee.ic.ac.uk/~wiser/message/index.html

11. J. H. Liu, et al., "An Air Quality Monitoring System for Urban Areas Based on the Technology of WSNs," International Journal on Smart Sensing and Intelligent Systems, Vol. 5, No. 1, 2012, pp. 191-214.

12. A. Bielsa, "Smart City Project in Serbia to Monitor Environmental Parameters by Public Transportation with Waspmote," 2012. http://www.libelium.com/smart_city_environmental_parameters_p ublic_transportation_waspmote

13. K. M. Tajne, et. al., "Monitoring of Air Pollution Using Wireless Sensors—A Case Study of Monitoring Air Pollution in Nagpur City," International Journal of Environmental Sciences, Vol. 2, No. 2, 2011, pp. 829-838.

14. R. V. Y. Prasad, et al., "Real Time Wireless Air Pollution Monitoring System," ICTACT Journal on Communication Technology: Special Issue on Next Generation Wireless Networks and Applications, Vol. 2, No. 2, 2011, pp. 370-375.

15. T. C. Keener, "WSNs for Monitoring Carbon Monoxide in Ambient Air," A&WMA International Speciality Conference Leapfrogging Opportunities for Air Quality Improvement, Xi'an, 10-14 May 2010, pp. 201-210.

16. J. Talašová, "Fuzzy Metody Vícekriteriálního Rozhodování," Olomouc, 2003, p. 179.

17. V. Voženílek, "Artificial Intelligence and GIS: Mutual Meeting and Passing," International Conference on Intelligent Networking And Collaborative Systems (INCOS 2009), 2009, pp. 279-284.

18. T. Přibylová, "Metody Rozhodování za Rizika a Jejich Použití v Ekonomické Praxi," Bachelor Thesis, University Pardubice, Pardubice, 2012. http://dspace.upce.cz /bitstream/10195 /46295/ 2/Pribylova T_Metody Rozhodovani_OS_2012.pdf

19. P. Tuček, V. Pászto and V. Voženílek, "Regular Use of Entropy for Studying Dissimilar Geographical Phenomena" Geografie, Vol. 114, No. 2, 2009. pp. 117-129.

20. H. Karl and A. Willig, "Protocols and Architectures for Wireless Sensor Networks," John Wiley & Sons, Ltd., Hoboken, 2005.

21. P. Holešinský, "Lokalizační Techniky Bezdrátových Senzorových Sítí," Electrorevue, Vol. 6, No. 12, 2009, pp. 661- 669.

22. J. Mizera, "Využití Senzorových Bezdrátových Sítí pro Monitorování Životního Prostředí," Bachelor Thesis, VUT Brno, Brno, 2011. http://www.vutbr.cz/www_base/zav_prace_soubor_verejne.php?file_id=42512

9

# CATHODIC PROTECTION REMOTE MONITORING BASED ON WIRELESS SENSOR NETWORK

**Mohammed Zeki Al-Faiz, Liqaa Saadi Mezher**

Department of Computer Engineering, AL-Nahrain University, Baghdad, Iraq

## ABSTRACT

Cathodic Protection system is an efficient system used for protecting the underground metal objects from corrosion. In this paper the use of Cathodic Protection (CP) system and how they can be developed to simulate corrosion control solution was illustrated. The aim of developing a Cathodic Protection system is to provide control over oil pipelines and to reduce the incidence of corrosion. The proposed system integrates the technology of wireless sensor Network (WSN) in order to collect potential data and to realize remote data transmission. In this system each WSN receives the data from the environment and forwards it to a Remote Terminal Unit (RTU). Then each RTU forwards it to its base station (BS). In this work Labview 2010 program was used, due to its high potentials. In addition it contains a Tool Kit that supports the wireless sensor network. In this simulation used many cases study to test and monitoring data and get optimum results, least time delay and high speed to prevent corrosion.

## KEYWORDS

Cathodic Protection System; Corrosion; Wireless Sensor Network; Remote Terminal Unit

# 1. INTRODUCTION

Cathodic protection is a method of corrosion control that can be applied to buried and submerged metallic structures. It is normally used in conjunction with coatings and can be considered as a secondary corrosion control technique. The primary corrosion control method on any given structure is normally a coating system which can be between 50% and 99% efficient depending upon age, type, method of installation, etc. A properly designed and maintained Cathodic protection system will take up the remainder resulting in a 100% efficient corrosion protection system [1].

There are two methods for Cathodic Protection these methods are Cathodic Current and Sacrificial Anode.

The principle of Cathodic protection is to connect an external anode to the metal to be protected and to pass a positive DC current between them so that the metal becomes Cathodic and does not corrode [2].

Cathodic protection has become a standard procedure for many structures such as underground storage tanks, pipelines, water storage tanks, ship hulls and interiors [3].

Corrosion is an electrochemical process that occurs when a current leaves a structure at the anode side, passes through an electrolyte (it is generally soil in buried metal pipeline), and reenters the structure at the cathode side, also known as oxidation [4].

The field of wireless sensor Network is gaining a rapid interest due to the application of smart low cost long battery life sensors. Usually sensor Network consist of the sensors as end devices, routers to choose the most appropriate path to an administrator host that is called the coordinator [5].

Many papers studied this problem like [6,7], in these paper Cathodic protection systems were discussed for water storage tanks, short distance, periodically tested, using voltmeter device to measure the voltage between the poles. In these systems, it is required the maintenance team to go down to the site in addition to high cost and time delay. While our paper proposed the wireless device use instead of voltmeter device. The proposed system helps the operator to observe the results of all sensors remotely from the monitor in the control room as well as reduce the overall cost and time delay.

In this paper, a new method was suggested handling the problem by CP for WSN and also used Labview simulator 2010 for emulated this method.

## 2. CATHODIC PROTECTION METHODOLOGY

Corrosion is an electrochemical process that involves passage of electrons from one substance, usually metallic, called the anode, to another substance called the cathode. It is at the anode that the oxidation reaction which is responsible for corrosion of the metal shows in the **Figure 1**(a) [8]. Show the following equations [3]:

$$2M \rightarrow 2M^{+} + 2e^{-} \quad (1)$$

where: M = metal; $M^+$ = soluble salt; $e^-$ = electron.

The basic process at an anodic site is the release of iron (Fe) from the steel surface into the environment and can be expressed as [8]:

$$\mathrm{Fe} \rightarrow \mathrm{Fe}^{+2} + 2e^{-} \quad (2)$$

During the process two electrons ($2e^-$) are generated which must be consumed by the environment (in aerated systems) and can be expressed as [1]:

$$4\mathrm{H}^{+} + \mathrm{O}_2 + 4e^{-} \rightarrow 2\mathrm{H}_2\mathrm{O} \quad (3)$$

This reaction produces free electrons, which pass within the metal to another site on the metal surface (the cathode), where it is consumed by the Cathodic reaction. In acid solutions the Cathodic reaction is [3]:

$$2\mathrm{H}^{+} + 2e^{-} \rightarrow \mathrm{H}_2 \quad (4)$$

In neutral solutions the Cathodic reaction involves the consumption of oxygen dissolved in the solution Show in **Figure 1**(b) [8]:

$$\mathrm{O}_2 + 2\mathrm{H}_2\mathrm{O} + 4e^{-} \rightarrow 4\mathrm{OH}^{-} \quad (5)$$

Corrosion thus occurs at the anode but not at the cathode (unless the metal of the cathode is attacked by alkali).

A summary of the above half reactions can be expressed as [1]:

$$2Fe + 2H_2O + O_2 \rightarrow 2Fe(OH)_2 \tag{6}$$

The term $Fe(OH)_2$ is iron oxide which can be oxidised to form the red-brown $Fe(OH)_3$ commonly referred to as rust.

Cathodic protection is an electrical method of preventing corrosion on metallic structures which are in electrolytes such as soil or water [6]. Therefore, it is necessary to control the corrosion of well casings for prevention of oil pipeline that can lead to environmental damage. Corrosion protection also ensures optimum economical depletion of oil reservoirs. The most effective method to decrease corrosion of well casings is the use of Cathodic Protection [9].

Advantage of Cathodic Protection methods is easy to install, no power source is required due to the fact that the protective current is generated by the electrochemical reaction between the metals, no externally supplied power is required, suitable for localized protection, minimum of maintenance required, no regulation is required, minimum of Cathodic interferences problems.The disadvantages of the Cathodic Protection include is small driving voltage available (the limited driving potential between the structure and the anode materials used), extremely small current available in higher resistivity electrolytes , not economically feasible to install or replace anodes on large or extensive existing structures, installation can be expensive if installed after construction [10].

# 3. THE SUGGESTED MONITORING SYSTEM

The Remote Monitoring of Cathodic protection system based on wireless Sensor model is one of the most effective, convenient and economical methods to prevent corrosion. Design changes can achieve the optimum performance of Cathodic protection systems, preventing the impact of corrosion.

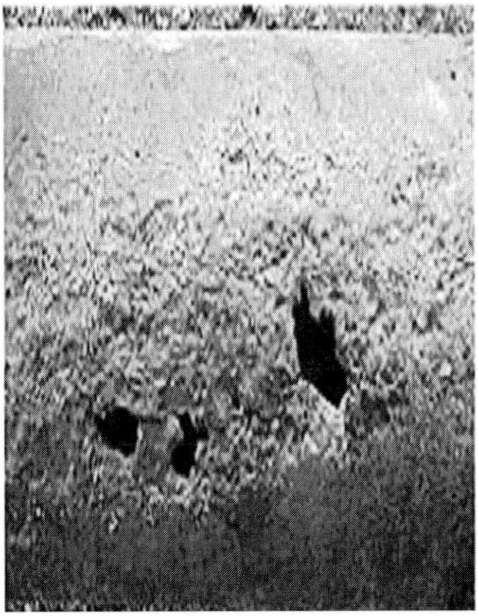

(a)

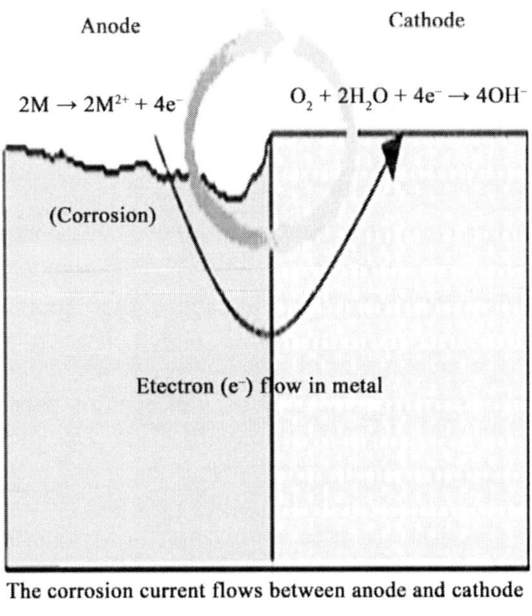


(b)

**Figure 1**. (a) Corrosion effects on pipelines; (b) Corrosion process.

The basic node of any WSN is the sensor node. The sensor node presented in this work consists of many basic parts such as: a sensing unit, a microcontroller, channel (transceiver), and battery as shown in **Figure 2**(a). The sensing unit is a sensor inputted to an analog to digital converter (ADC) block embedded in the microcontroller. The communication unit is transceiver suitable for converting the signals of the microcontroller to the signal.

In this paper, we used a hierarchical architecture consisting of a sensor node, a gateway (RTU), and a base station. Sensor node that is responsible for monitoring the physical environment placed at the lowest level of the hierarchy. The designed network architecture of this work is illustrated in **Figure 2**(b).

Wireless Sensor Network was used instead of the voltmeter for CP, where connected these devices with the parties to the anode and cathode of Cathodic protection, and then installed these devices the ground at a distance of one meter from the Oil pipeline, and the distance between two Wireless Sensor 2 km. The work of these devices is to collect data in the form of analog signal and convert them into digital signal and sent to the RTU and after gathering information from all wireless sensor devices, then all RTU stations send them to the base station to monitor and test data and give the result on the monitoring form. In **Table 1** the Remote Monitoring used many technologies.

# 4. PROBLEM SECTION

The oil pipelines exposed continuously to corrosion or rust due to external factors from environment and to reduce this problem. Cathodic protection methods are used to prevent corrosion. These methods are preventing access of electrolyte, reversing the flow of electrons, corrosion resistance alloys like stainless steel. The method is widely used in underground and seawater structures.

The principle of Cathodic protection consists in the connection of an external anode to the metal to be protected and the passing of an electric current so that all areas of the metal become Cathodic and therefore not corrode. In electrochemical terms, the electrical potential between the metal and the electrolyte solution with which it is in contact is lowered to a value at which corroding anodic reactions are stifled and only Cathodic reactions can take place [12].

Both anode and cathode connected with wireless sensors. Anode connected with the positive from wireless and cathode connected with the negative from wireless sensor. Then the wireless sensor collected the data from environment through the difference voltage between the anode and cathode and sends it to remote terminal unit. Then the remote terminal unit forwards data to base station (control room) and here the process of the control data examine and make it update each time period according to the location and significance.

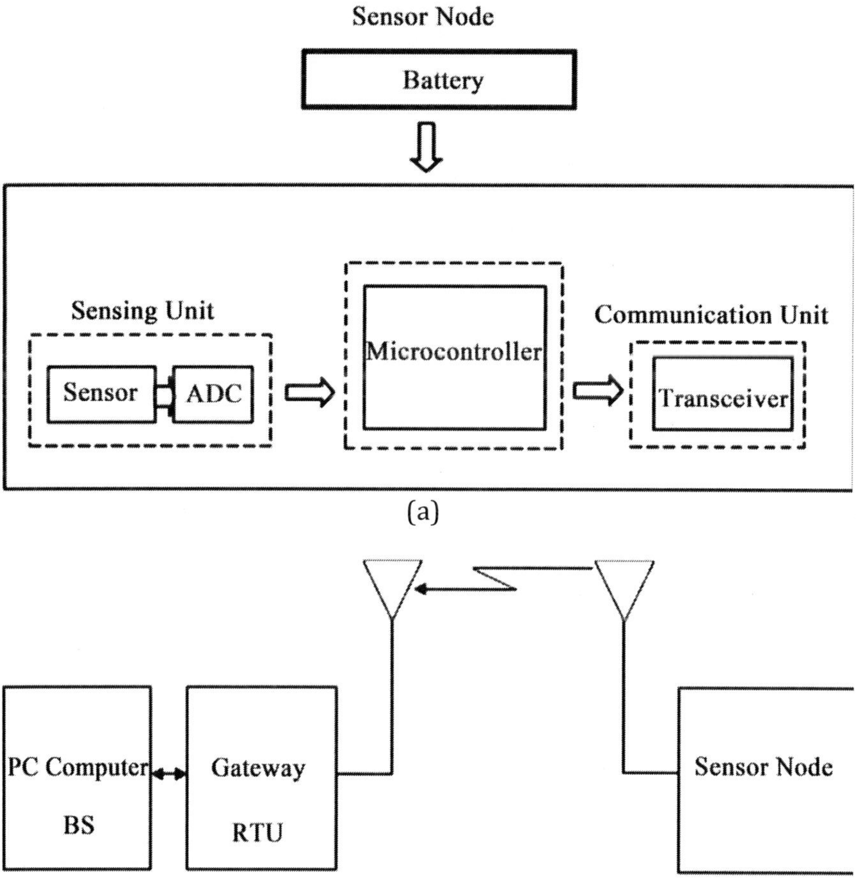


(b)

**Figure 2**. (a) Block diagram of Wireless Sensor [11]; b) Block diagram of the proposed network architecture designed for data acquisition [11].

**Table 1**. Comparison of Cathodic Protection Remote Monitoring Technologies.

| | SCADA-Based System | Satellite-Based System | Cell phone-Based System |
|---|---|---|---|
| **Complexity** | High | High | Low |
| **Advantages** | —Takes advantage of existing wire-less system<br>—can take readings at any time<br>—can control | —Monitor CP system virtually any time, anywhere in the world through the internet | —Low installation cost<br>—can easily be installed<br>—can take readings at any time |
| **Disadvantages** | —Signal path must be properly planned for adequate transmission Signals must be line of site(LOS)<br>—Signals are sometimes lost if there are impediments to LOS | —Charges fee/reading/location<br>—Line of site limitations | —Cost of cell phone services<br>—Cell phone signals are sometimes lost in certain locations |

# 5. CASE STUDY

As an example a Cathodic protection methodology is considered for the Oil pipelines. In this method the cathode and anode are buried under the ground in a distance of two meters. Both cathode and anode are connected to the wireless sensor. This sensor is installed in one meter from the Oil pipeline and the distance between each two sensors of WSN is two kilometers. The data are collected from the environment by wireless sensor and sent to the RTU. Each RTU collection data from range about 50 kilometers and then forward to base Station. The distance between RTU and Base Station is 100 kilometers.

In this simple example nine wireless sensor devices, three RTU and Base Station. Each RTU included three wireless sensors. This Wireless Sensor was collected data from environment and sends it to RTU and then the RTU forwards data to base station in order to conduct testing and monitoring them by using Labview simulation program.

# 6. SIMULATION & RESULT

The proposed algorithm was simulated in a Labview 2010 simulation program. Depended on Standard Colamer Electric (SCE) to testing the data Acquisition.

## 6.1. Sensor Side of WSN

The flow chart of the sensing program is shown in **Figure 3**. **Figure 4** shows the measurement sequence. First the sensor side is initialized by setting the service port number and defining the IP Address of the gateway (RTU). After that it starts a timer for the TCP listen block. This block obtains TCP packets from gateway (RTU) in the range of around 50 km. Furthermore this operation depends on the IP address obtained in the beginning of the process. The duration of this step is chosen to be (6 sec.). Each sensor sends its data to the RTU within 2 seconds. After 6 seconds the communication between the RTU and sensors is terminated. Next new sensors start to communicate with the other RTUs. After (6 sec.) the RTU will close the TCP connection. Otherwise it resets the timer and then it uses TCP write block command to write the forward packet data to the RTU. The Sensor program continues to send packets to the gateway unless a manual stop action is initiated from the Sensor.

## 6.2. Gateway Side (RTU)

The flow chart of the gateway program is shown in **Figure 5**. Also **Figure 6** shows the measurement sequence. It also starts by initialized this side by setting the port number and the IP address of the RTU. Then the program starts a timer for TCP listen block. This block search for the end nodes (in this case the wireless sensor) and obtain the values of the signals from it. Then the founded sensors by the TCP listen are connected to the RTU. After that the founded sensors send their values to the RTU. The maximum time allowed for this step is 18 seconds. However, if the RTU side did not receive the signal value within this period it will close the TCP connection.

In general at the end each timer the obtained values by RTU are organized in a file and send back to the BS. The BS read the received information and the following test is made:

1) If the value obtained from the sensor is (−1.70 to −0.85) V, this means that the pipeline is plated, protected and activated as shown in **Figure 7**.

2) If the value obtained from the sensor is in the range (−0.85 to −0.25) V, this means that the pipe line is in an alarm status. Furthermore a timer is set for a specified period and a manual action must be made by the user. In addition no further readings are taken from that sensor until an action is made. The timer period is set according to the level of risk in the oil pipeline.

- If the value obtained from the sensor is in the range (−0.85 to −0.65) V, this means that the pipe line is plated, unprotected and it is in the danger of exhibition of corrosion as shown in **Figure 7**.1.

- If the value obtained from the sensor is in the range (−0.65 to −0.45) V, then the pipe line in lightly plated, unprotected (a reason for this either that the apart of the anode exhibit a corrosion or is disconnected) and it is in the danger of exhibition of corrosion as shown in **Figure 7**.2.

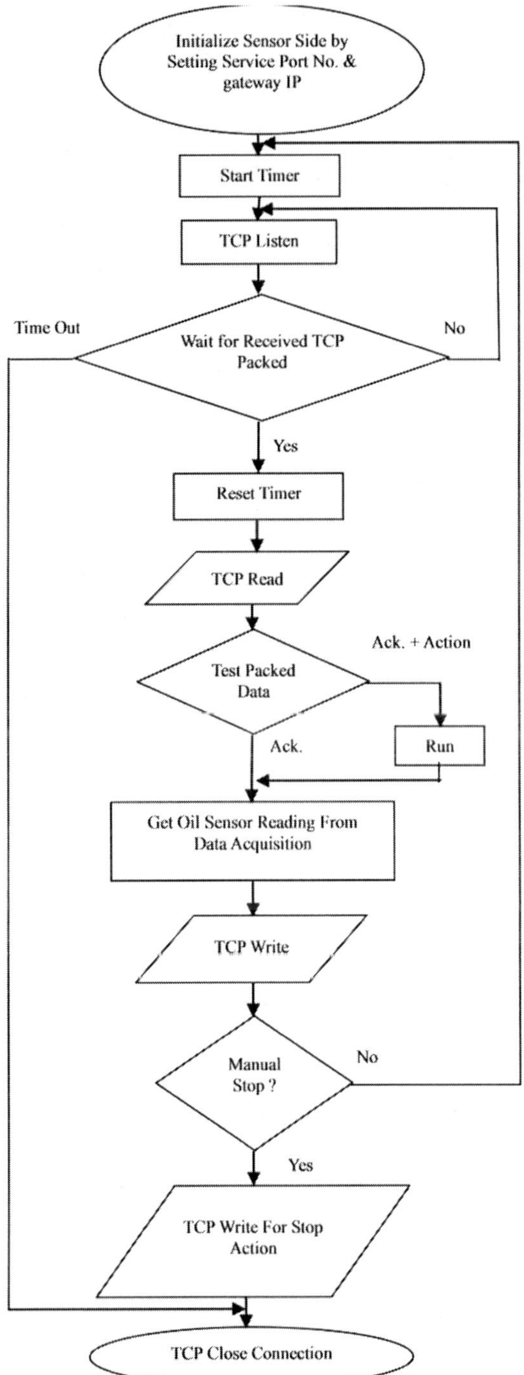


**Figure 3** flow chart of Sensor side program.

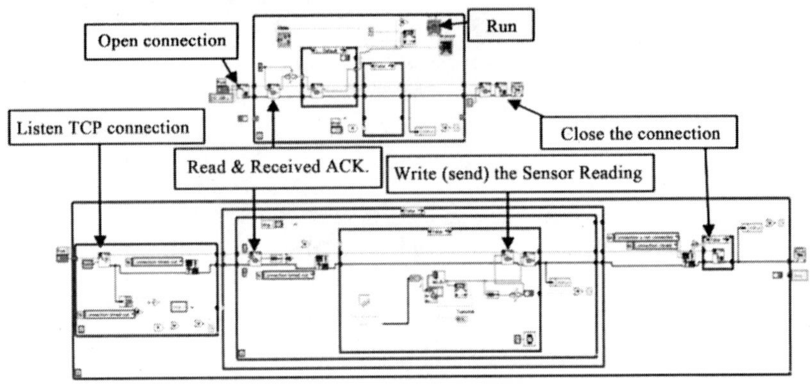


**Figure 4**. Lab VIEW Block diagram of the Sensor side.

**Figure 5**. Flow chart of the gateway (RTU) side program.

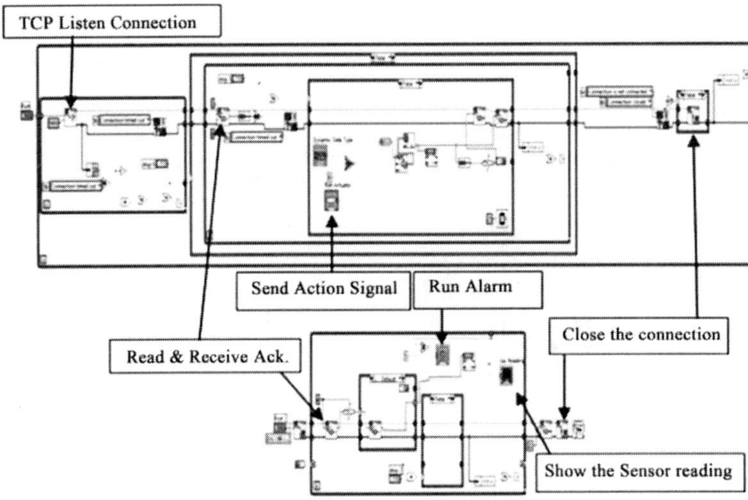


**Figure 6**. Block diagram of the coordinator side.

- If the value obtained from the sensor is in the range (−0.45 to −0.25) V, this means that the pipe line is not plated, unprotected (a reason for this either that the anode exhibit a corrosion or is disconnected) and it is in the danger of exhibition of corrosion as show **Figure** 7.3.

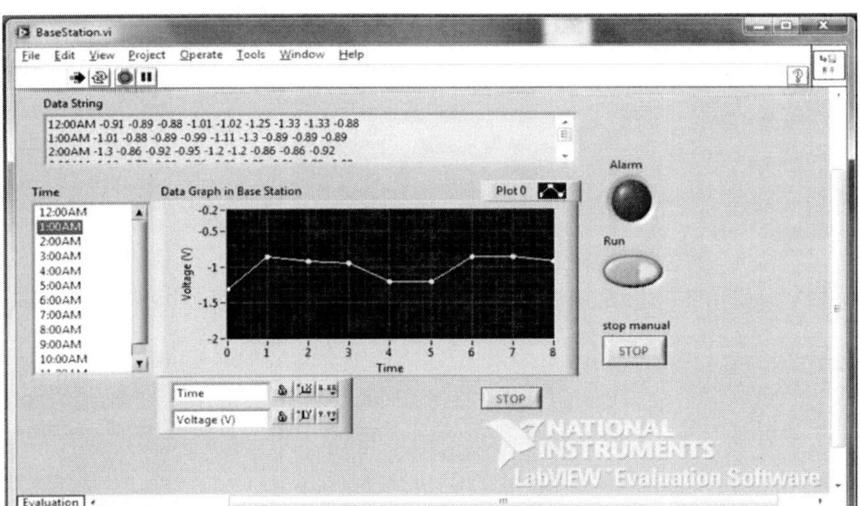


**Figure 7**. Normal case study in BS Front Panel.

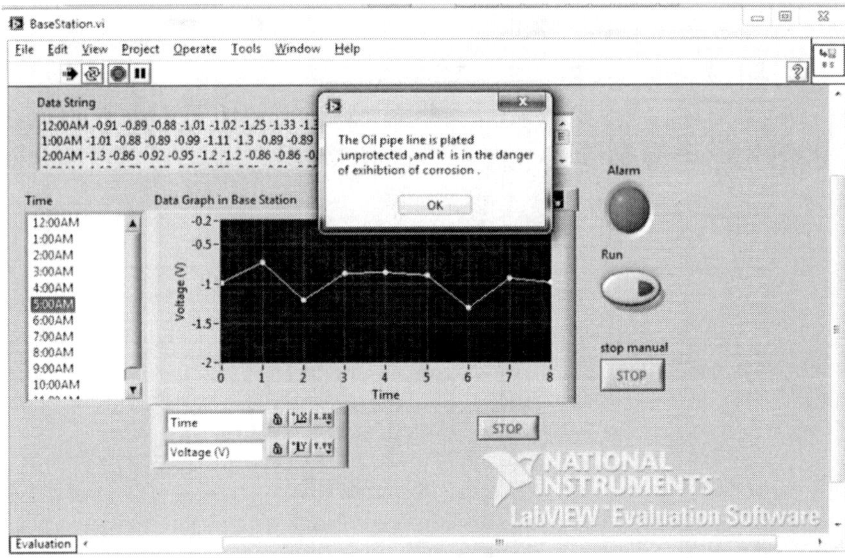


**Figure 7**.1. Alarm case study 1 in BS Front Panel.

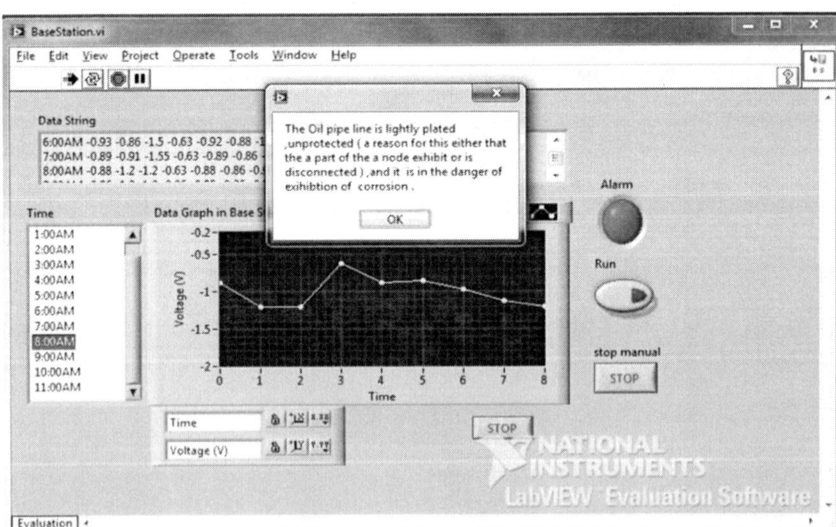


**Figure 7**.2. Alarm case study 2 in BS Front Panel.

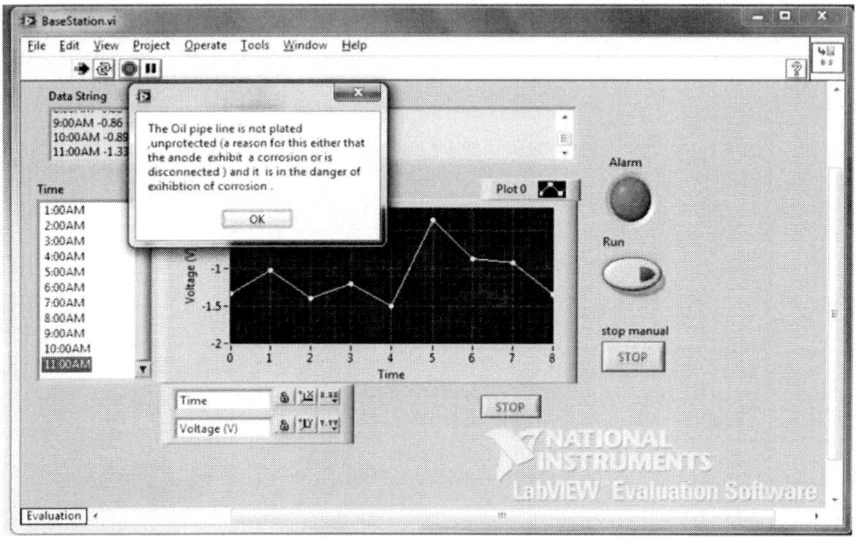


**Figure 7**.3. Alarm case study 3 in BS Front Panel.

In **Table 2** Data from all WSNs with Update, The update of the sensor values to the base station is made every hour. **Figure 6** shows the Block diagram for the gateway side program.

# 7. CONCLUSIONS

The detection system based on Cathodic protection. In this paper was enhanced detection accuracy of oil pipeline to a great extent. This system adopts multi-wireless sensor network, it has a very great superiority compare with the Voltmeter device connected directly with Oil Pipeline. It can detect the corrosion of oil pipeline precisely and reliably.

To conclude, a WSN is suggested to avoid corrosion problem in oil pipeline in state of the common CP method. In addition, a new simulation method of WSN is presented. Labview 2010 program is used as the tool to build the simulation environment. The results of a case study have been illustrated. The simulation results show that this method has least time delay, high speed, and high quality of service.

**Table 2**. Base station data from all Wireless Sensor Network with update (Voltage in V).

| Time | RTU1 | | | RTU2 | | | RTU3 | | |
|---|---|---|---|---|---|---|---|---|---|
| | WSN1 | WSN2 | WSN3 | WSN4 | WSN5 | WSN6 | WSN7 | WSN8 | WSN9 |
| 12:00AM | −0.91 | −0.89 | −0.88 | −1.01 | −1.02 | −1.25 | −1.33 | −1.33 | −0.88 |
| 1:00AM | −1.01 | −0.88 | −0.89 | −0.99 | −1.11 | −1.3 | −0.89 | −0.89 | −0.89 |
| 2:00AM | −1.3 | −0.86 | −0.92 | −0.95 | −1.2 | −1.2 | −0.86 | −0.86 | −0.92 |
| 3:00AM | −1.12 | −0.73 | −0.99 | −0.86 | −0.98 | −0.95 | −0.91 | −0.88 | −0.99 |
| 4:00AM | −0.99 | −0.73 | −1.3 | −0.86 | −0.91 | −0.92 | −1.01 | −0.89 | −0.99 |
| 5:00AM | −0.98 | −0.73 | −1.2 | −0.87 | −0.86 | −0.9 | −1.3 | −0.93 | −0.98 |
| 6:00AM | −0.93 | −0.86 | −1.5 | **−0.63** | −0.92 | −0.88 | −1.12 | −0.98 | −1.35 |
| 7:00AM | −0.89 | −0.91 | −1.55 | **−0.63** | −0.89 | −0.86 | −0.99 | −0.99 | −1.55 |
| 8:00AM | −0.88 | −1.2 | −1.2 | **−0.63** | −0.88 | −0.86 | −0.97 | −1.12 | −1.2 |
| 9:00AM | −0.86 | −1.2 | −1.2 | −0.86 | −0.95 | **−0.36** | −0.95 | −1.3 | −1.4 |
| 10:00AM | −0.89 | −1.11 | −1.3 | −0.92 | −1.2 | **−0.36** | −0.89 | −1.01 | −1.3 |
| 11:00AM | −1.33 | −1.02 | −1.4 | −1.2 | −1.5 | **−0.36** | −0.87 | −0.92 | −1.35 |

## REFERENCES

1. P. R. Roberge, "Handbook of Corrosion Engineering," McGraw-Hill PY, New York, 2000.

2. N. Summers, "Remote Monitoring of Pipeline Cathodic Protection Systems," Petromin Pipeline, April-June 2009.

3. M. Hafiz, "Modeling of Pipeline Corrosion Control by Cathodic Protection," Ph.D. Thesis, AL-Technology University, Baghdad, 2006.

4. S. Gundogdu and O. Sahin, "E.M.I Effects of Cathodic Protection on Electromagnetic Flowmeters," Sensors, Vol. 7, No. 1, 2007, pp. 75-83.

5. M. Al-Somaidai and O. Alsaydia, "Remote Monitoring and Controlling of Gas Sensors Using VPN Connection," The First International Conference on Future Communication Networks ICFCN, Baghdad, 2-5 April 2012, pp. 84-87.

6. J. Bushman, "Corrosion and Cathodic Protection Theory," Bushman & Associates Inc., Medina, 2010.

7. A. Kumar and L. D. Stephenson, "Comparison of Wireless Technologies for Remote Monitoring of Cathodic Protection Systems," US Army Corps of Engineers, Engineer Research and Development Center Construction Engineering Research Laboratory, Champaign, 2007.

8. L. Ting and Z. Xiang, "Optimization Of Remote Monitoring of Cathodic Protection Power Supply System Base on Four-Tier Structure and B/S Model," The 6th International Conference Computer Science & Education (ICCSE), Singapore City, 3-5 August 2011, pp. 1346-1350.

9. M. Parsa, S. Allahkaram and A. Ghobadi, "Simulation of Cathodic Protection Potential Distributions on Oil Well Casings," Journal of Petroleum Science and Engineering, Vol. 72, No. 3-4, 2010, pp. 215-219.

10. J. Bushman, "Galvanic Anode Cathodic Protection System Design," Bushman & Associates Inc., Medina, 2010.

11. A. Abed, "Design and Implementation of a Wireless Sensor/Actuator Network for the Development of an Industrial Wireless DCS System," Ph.D. Thesis, AL-Basra University, Basra, 2012.

12. J. T. Al-Haidary, M. H. Hafiz and Y. M. Abdu Al-Sahib, "Galvanic Cathodic Protection Evaluation of a Steel Pipe in Iraqi Soil," Journal of Medical Engineering & Technology, Vol. 29, No. 9, 2011, p. 1673.

10

# REFRIGERATED FRUIT STORAGE MONITORING COMBINING TWO DIFFERENT WIRELESS SENSING TECHNOLOGIES: RFID AND WSN

**Ricardo Badia-Melis [1,*], Luis Ruiz-Garcia [1], Javier Garcia-Hierro [2] and Jose I. Robla Villalba [2]**

[1]Departamento de Ingeniería Agroforestal, ETSI Agrónomos, Universidad Politécnica de Madrid, 28040 Madrid, Spain

[2]Sensors Technology Laboratory, Centro Nacional de Investigaciones Metalúrgicas, Consejo Superior de Investigaciones Científicas (CENIM-CSIC), Avenida Gregorio del Amo 8, 28040 Madrid, Spain

## ABSTRACT

Every day, millions of tons of temperature-sensitive goods are produced, transported, stored or distributed worldwide, thus making their temperature and humidity control essential. Quality control and monitoring of goods during the cold chain is an increasing concern for producers, suppliers, logistic decision makers and consumers. In this paper we present the results of a combination of RFID and WSN devices in a set of studies performed in three commercial wholesale chambers of 1848 $m^3$ with different set points and products. Up to 90 semi-passive RFID temperature loggers were installed simultaneously together with seven motes, during one week in each chamber. 3D temperature mapping charts were obtained and also the psychrometric data model from ASABE was implemented for the calculation of enthalpy changes and the absolute water content of air. Thus thank to the feedback of data, between RFID and WSN it is possible to estimate energy consumption in the cold room, water loss from the products and detect any condensation over the stored commodities.


## KEYWORDS

cold chain; logistics; post-harvest; wireless sensor networks; RFID


## 1. INTRODUCTION

The production, storage, distribution and transport of cold sensitive products, occurs daily around the world. For all these products the control of temperature is essential [1]. The term "cold chain" describes the series of interdependent equipment and processes employed to ensure the temperature preservation of perishables and other temperature-controlled products from the production to the consumption end in a safe, wholesome, and good quality state [2].The inadequate management of this supply chain can reduce the acceptable quality limit of fresh fruit and vegetables (FFVs) making approximately one-third of these products be thrown away [3]. Thus, the major challenge is to ensure a continuous cold chain from producer to consumer in order to guarantee the prime condition of goods [4]. These products may be perishable items like fruit, vegetables, flowers, fish, meat and dairy products or medical products like drugs, blood, vaccines, organs, plasma and tissues. The properties of all of them are affected by temperature changes. In case of deviations from the optimal cold chain conditions, the shelf life of the sensitive products will be reduced, yet the consequences of this shelf life drop are only physically visible in the advanced stages of the supply chain in many occasions. To understand this variation is possible by a remote monitoring [5].

The quality of these products might change rapidly due to inadequate temperature and relative humidity conditions during transport and storage. Temperature variations can occur during warehousing, handling and transportation. Inadequate temperature is second on the list of factors causing foodborne illnesses, surpassed only by the presence of initial microflora in foods [6]. The U.N. FAO estimates that each year, approximately one-third of all food produced for human consumption in the world is lost or wasted [7]. Thus, studying and analyzing temperature gradient data inside refrigeration rooms, containers and trucks is a primary concern for the industry. Any temperature disturbance can undermine the efforts of the whole chain [8].

It is not easy to maintain appropriate conditions over the whole chain, and negligence or mishandling in the logistics of perishable food products is very common, including goods poorly or excessive cooled. Jedermann [9] reported many examples where the inadequate management during temperature control usually leads to losses in the food chain (post-harvest, distribution and at home).

Apart from temperature, water loss is one of the main causes of deterioration that reduces the marketability of fresh fruits and vegetables. Transpiration is the loss of moisture from living tissues. Most weight loss of stored fruit is caused by this process. Relative humidity (RH), temperature (T) of the product, temperature of the surrounding atmosphere, and air velocity all affect the amount of water lost from perishable food products. Free water or condensation is also a problem as it encourages microbial infection and growth, and it can also reduce the strength of packaging materials [10,11].

Wireless technologies have been considered key technological enablers, whereby intelligent tags are attached to an item, within a network with wireless links, which can transmit some physical parameters such as temperature or humidity in addition with other information like position or movement [12].

Some studies have demonstrated that a proper control over the cold chain using RFID loggers can lead to reduced waste of products by using the "First Expiring First Out", rather than the classical "First In First Out" paradigm, and also have implemented prediction temperature methods in order to improve the estimation of shelf life [13]. RFID tags are not limited to identification or temperature and RH reading tasks, and recent advancements have produced RFID units able to identify volatile chemicals such as ethylene or ammonia [14].

Specialized Wireless Sensor Network (WSN) monitoring devices can revolutionize the shipping and handling of a wide range of perishable products by providing suppliers and distributors with continuous and accurate readings throughout the distribution process. In this framework, WSN was developed as a very promising technology due to its low energy consumption and advanced networking capabilities. The potential for monitoring with both technologies, WSN and RFID, has been suggested by several studies; however there is still a lack of implementation and experimentation in real environments, and no study combines the advantages of using both at the same time [15,16,17,18,19].

The paper is focused on the analyses of the performance of semi-passive RFID loggers and WSN motes, in order to enable an economical solution for the spatial profiling of refrigerated chambers, with a high number of loggers. The aim of this work is to demonstrate the compatibility and feedback in a combination of the RFID and WSN in order to improve the refrigerated storage of perishable food products. The authors also want to provide some recommendations to improve current wireless devices and some management protocols or rules for the usage of the cold storage rooms.

# 2. MATERIALS AND METHODS

## 2.1. Refrigerated Chamber

Experiments were performed in three commercial wholesale refrigerated stores, numbered 11, 29 and 40 and each of which has a volume at of 26 m × 6 m × 12 m = 1848 $m^3$, with an on/off glycol cooling system and insulated walls built of foam sandwiched between two layers of corrugated plate (total wall thickness is 0.16 m). The set point experimentation times were different in each one (see Table 1). Each chamber has a common pre-chamber space where a devoted sensor is placed. Therefore, two different ambient conditions occur: pre-chamber (right outside the chamber) and the inside chamber with a well-known set point (see Table 1).

**Table 1.** Experimental conditions.

| Cold Store Number | Set Point (°C) | Experimentation Time (days) | Dates |
|---|---|---|---|
| 11 | 8 | 13 | 7–19 July |
| 29 | 7 | 8 | 20–27 July |
| 40 | 14 | 4 | 28–31 July |

Chamber number 11 was divided into two equal size sections by a wall with a door that remained open all the time. The dimensions of the door were the same as the main door to the exterior of the whole room, *i.e.*, it was big enough to allow a forklift with a big pallet to pass. The room was provided with two cooling systems, one per each section of the chamber. Inside were two shelves on either side of the room with three heights, where the products were stored. When the temperature of the room reached 8 °C the refrigerator was triggered to cool down to 6.5 °C. A scheme of the chambers can be seen in Figure 1.

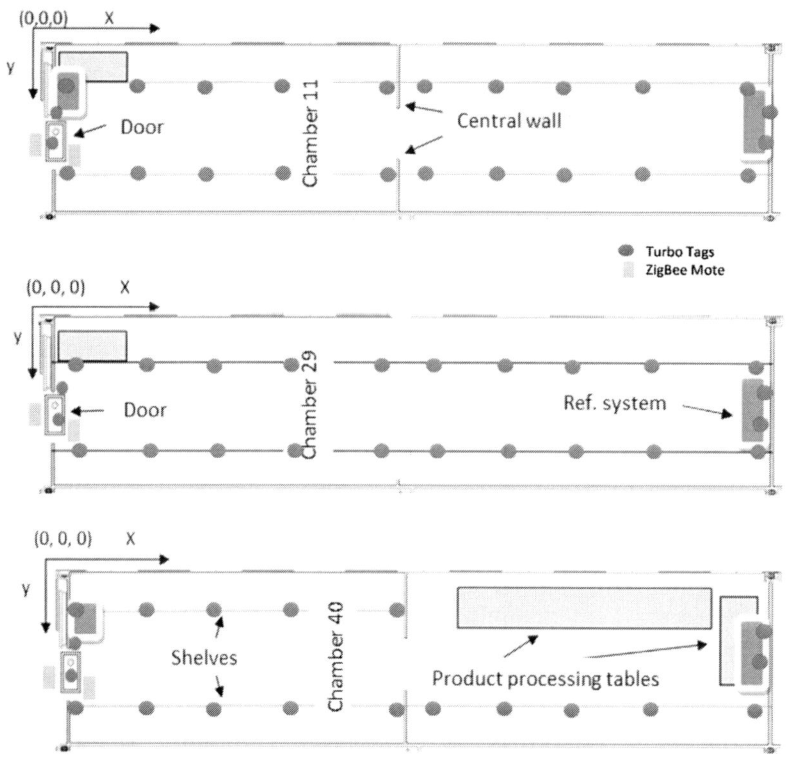


**Figure 1.** Scheme of the cold rooms and sensor distribution.

Chamber number 29 had identical dimensions but with no separating wall in the middle, and the temperature was set at 7 °C and dropped to 5.5 °C. Both these rooms were designated to store mostly citric fruits and different varieties of nuts.

The last chamber (number 40) was divided into two sections separated by a wall with a semiautomatic blind door; the first section was designated for the storage of vegetables and the other section was designated for fruit and vegetable packaging.

## 2.2. WSN Motes

Two ZigBee/IEEE 802.15.4 motes (transmitters) and one base station (receiver) were used. One mote was installed outside the chamber, close to the door and the other one inside, at the other side of the wall. Sampling rate was set to 180 s.

These motes were manufactured by Crossbow® (Milpitas, CA, USA ) and they have a microcontroller board (IRIS, Milpitas, CA, USA) together with an independent transducer board (MTS400, Andover, MA, USA) attached by means of a 52 pin connector.Its processor & radio platform is a XM2110CA, based on the Atmel ATmega1281. The RF power was configured to 3 dBm (60% higher than previous MICA Motes). Power was supplied by two AA lithium batteries.

The MTS400 board hosts a variety of sensors: temperature and relative humidity (Sensirion SHT, Staefa, Switzerland), barometric pressure and temperature (Intersema MS5534B, Hampton, VA, USA), light intensity (TAOS TSL2550D, Plano, TX, USA) and a two-axis accelerometer (ADXL202JE, Norwood, MA, USA). A laptop computer is used as the receiver, and it communicates with the nodes through a Micaz mounted on the MIB520 ZigBee/USB gateway board; this device also provides a USB programming interface. In this study only the data from the Sensirion and Intersema units was used.

The Sensirion SHT is a single-chip relative humidity and temperature multi-sensor module that delivers a calibrated digital output. Each SHT is individually calibrated in a precision humidity chamber. The calibration coefficients are programmed into the One Time Programmable (OTP) memory. These coefficients are used internally during measurements to calibrate the signals from the sensors.

For temperatures significantly different from 25 °C, and according to the manufacturer recommends performing humidity and temperature compensation, using Equations (1) and (2):

$$RH_{linear} = (-4) + 0.0405 \times SO_{RH} + (-2.8 \times 10^{-6}) \times SO_{RH}^2 \quad (1)$$

where $SO_{RH}$ = Sensor Output Relative Humidity.

$$RH_{true} = (T\ ^\circ C - 25) \times (0.01 + 0.00008 \times SO_{RH}) + RH_{linear} \quad (2)$$

The MS5534B is a SMD-hybrid device including a piezoresistive pressure sensor and an ADC-Interface IC. It provides a 16 bit data word from a pressure and temperature (−40 to +125 °C) dependent voltage. Additionally the module contains six readable coefficients for a highly accurate software calibration of the sensor. MS5534B is a low power, low voltage device with automatic power down (ON/OFF) switching. A 3-wire interface is used for all communications with a microcontroller. Sensor packaging options are plastic or metal caps.

## 2.3. Semi-Passive RFID Tags

RFID tags can be active, passive, or semi-passive. Passive and semi-passive RFID send their data by reflection or modulation of the electromagnetic field that was emitted by the reader. The battery of semi-passive RFID is only used to power the sensor and recording logic [1]. In this study we used semi-passive tags.

Up to 90 RFID semi-passive loggers were installed in each chamber, the tags are distributed in four layers or heights; each layer is as shown in Figure 1. The tags were manufactured by Sealed Air (New York, NJ, USA), and two different types were used, a Turbo Tag T700 (−30 °C to +55 °C) and the T702-B (−70 °C to +80 °C). Each tag operates in the 13.56 (ISO 15693-3 compliant) MHz band, has the size of a credit card, is accurate to within ±0.5 °C, and both have KSW Microtec (Dresden, Germany) temperature sensors and can store up to 702 temperature measurements.

## 2.4. Data Analysis

In the particular case of the WSN system, since its task consists of sending data to a gateway in a continuous flow, a specialized Matlab program has been developed for assessing the percentage of lost packets (%) in transmission, by means of computing the number of multiple sending failures for a given sample rate (SR). A multiple failure of "m" messages occurs whenever the elapsed time between two messages lies between 1.5 × m × SR and 2.5 × m × SR. For example, with a sample rate of 11 s, a single failure (m = 1) occurs whenever the time period between consecutives packets is longer than 16.5 s (1.5 × 1 × 11) and shorter than 27.5 s (2.5 × 1 × 11). The total number of lost packets is computed based on the frequency of each failure type. Accordingly, the total percentage of lost packets is calculated as the ratio between the total number of lost packets and the number of sent packets.

The standard error (SE) associated to the ratio of lost packets is computed based on a binomial distribution as expressed in Equation (3), where n is the total number of packets sent, and p is the ratio of lost packets in the experiment:

$$SE = \sqrt{\frac{p(1-p)}{n}} \tag{3}$$

A second dedicated Matlab code was used for plotting 3D temperature gradients. This program makes use of linear spatial interpolation in order to obtain 3D representations of normalized temperatures and variances inside the chambers. Due to the significant variation of external conditions in the Spanish summer, temperature analysis make use of a normalized temperature difference (ΔTn), which is computed with respect to the set point and to the outside temperatures (see Equation (4)). This value gives a normalized measure with respect to the varying ambient conditions of the experiments:

$$\Delta T_n = \sqrt{\frac{T_m - T_s}{T_e - T_s}} \tag{4}$$

where $\Delta T_n$ (dimensionless) stands for the normalized temperature difference, $T_m$ is the average temperature value (°C) of each RFID logger, $T_s$ is the temperature set point value (°C), and $T_e$ refers to the average outside temperature (°C).

Finally, a 3D plot of indoor temperature variance with regard to outdoor temperature variance is recorded in order to represent the indoor temperature variability corrected with regard to changes in ambient experimental conditions (see Equation (5)):

$$V_n = \frac{V_i}{V_e} \tag{5}$$

where $V_n$ is the normalized indoor variance for experiment, $V_i$ is the indoor temperature variance of each logger, and $V_e$ is the outside variance. The indoor variability is expected to be lower than the pre-chamber variability, this value will give us an estimation of the differences in the temperatures in the two ambient; the closer to the unit the normalized indoor variance is, the closer the two variances are. This can provide an idea of the influence of the outdoor temperatures without taking into account the set point, therefore it is possible to evaluate the transit of the staff and the isolation of the cold room itself, while in the normalized temperature difference the cooling unit appears as a factor in the equation. The complete set of data is included for each experiment:

## 2.5. Psychrometric Data

The ASAE D271.2 standard, defined in April 1979 and reviewed in 2006, is used for computing the psychrometric properties of air [20]. Equations (6)–(8) and Table 2 enable the calculation of all psychrometric data of air whenever two independent psychrometric properties of an air-water vapor mixture are known in addition to the atmospheric pressure:

$$Ps = R * e^{\frac{A+B*T+C*T^2+D*T^3+E*T^4}{F*T-G*T^2}} \tag{6}$$

273.16 K ≤ T ≤ 533.16 K. $T$ = Temperature (K), $Ps$ = Saturation vapour pressure (Pa) [20].

$$Pv = Ps\frac{RH}{100} \tag{7}$$

$Pv$ = Vapor pressure (Pa) [20].

$$H = \frac{0.6219 * Pv}{Patm - Pv} \tag{8}$$

$H$ = Absolute humidity (g/kg dry air), $P_{atm}$ = Atmospheric pressure (Pa) [20].

**Table 2.** Coefficients used to compute the psychrometric data, according to Equation (7) [20].

| | |
|---|---|
| R = 22.105.649.25 | $D = 0.12558 \times 10^{-3}$ |
| A = −27.405.526 | $E = -0.48502 \times 10^{-7}$ |
| B = 97.5413 | F = 4.34903 |
| C = −0.146244 | $G = 0.39381 \times 10^{-2}$ |

# 3. RESULTS AND DISCUSSION

The presentation of the results in this work is divided into three parts. The first is the percentage of lost packets in the WSN transmission, which is key knowledge since is telling how much information is lost in translation, and it is saying that probably there should be a better transmission, the reasons are either a reading range problem, volumes of data are too big or there is some physical limitation, e.g. a massive amount of loggers, as Ruiz-Garcia & Lunadei address in their article [1], there are certain challenges and limitations that hinder a perfect data flow. That is another reason why RFID shows good complementarity with WSN, and the data loss can be addressed with the RFID tags.

The second part is about the temperature distribution, and as it was only possible to carry out this measurement due to the high volume of data generated from the RFID tags; it allows creating a uniform mesh of temperature nodes with many data points which is necessary for the 3D representation, in this case this big volume was not a limitation, but when there is a good understanding of the environment, certain spots of the cold room can be monitored with temperature estimation methods and bypass the actual sensors, as was successfully shown by other authors [21,22,23].

The third part is the psychrometric representation, where both the humidity sensor like the temperature sensor (in the WSN nodes) are gathered in the Sensirion SHT unit. This graphics are meant to compare the air inside and outside the cold rooms close to the door.

## 3.1. Percentage of Lost Packets

The 3 dBm of RF power used by the motes was enough to enable sending data outside the chamber. More than 98% of the packets sent were able to cross the walls (see Table 3). Only 1.25% of signals were lost in the worst case (Chamber 11). In previous experiments [24] under very similar conditions, the average of lost packages was 1.92% ± 0.07%, in that case the gateway was located in the same room as the mote and in our experiment one of the motes is across the isolating door of the room for better performance.

**Table 3.** Percentage of data lost packets and standard error during the experiments.

| Cold Store Number | Mote Inside (Data) | Mote Outside (Data) |
|---|---|---|
| 11 | 0.26 ± 0.01 (6117) | 1.25 ± 0.01 (6007) |
| 29 | 0.24 ± 0.02 (3244) | 0.89 ± 0.02 (3223) |
| 40 | 0.21 ± 0.02 (1846) | 0.70 ± 0.02 (1826) |

## 3.2. 3D Plots

Before starting with the 3D views, it is important to take a look on the absolute temperatures, which will give the reader a better understanding of the following Figure 3 and Figure 4. Figure 2a shows the cooling cycles inside the cold room, recorded by the sensors in the WSN mote and the RFID tag located near the mote. Accordingly with the normalized temperature one can appreciate that the temperature from the Turbo tag rarely surpasses the set point, and that gives a negative normalized temperature. In Figure 2b the day and night cycles are shown and the differences between the maximum and minimum values are around 5 °C and are bigger than inside the room (around 3 °C), which gives a normalized indoor variance between 0 and 1.

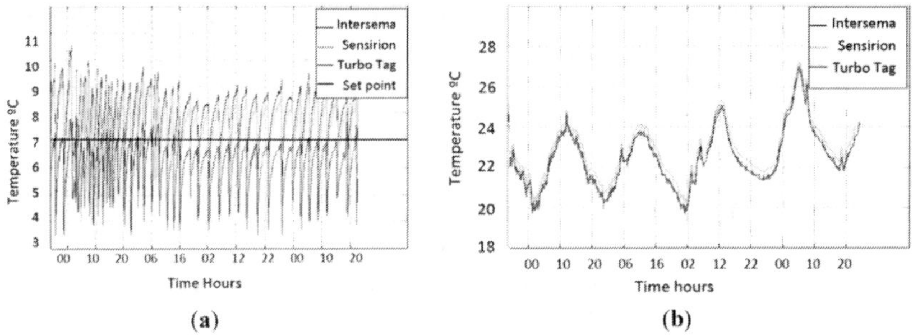


**Figure 2.** (**a**) Absolute temperature inside chamber 29; (**b**) Absolute temperature pre-chamber 29.

Figure 3 and Figure 4 show a summary of the 3D views corresponding to temperature behavior for Chambers 11 and 29. Both the normalized indoor variance ($\Delta T_n$) and normalized temperature variance are represented. In both cases mean temperatures were close to the set point, which means almost 0 normalized indoor variance ($\Delta T_n = 0$). The highest differences in room 11 were registered by the sensors located in the ceiling, where in the back part of the room it can reach a normalized temperature difference (NTD) of 0.01 points and in a spot in the front part it goes to −0.05 points. In Figure 3, in Chamber 11 one can appreciate how the temperatures are closer to the set point in the back part of the room with an average of −0.01 than in the front part with a lower average −0.03, the front part presents the most different temperature from the set point despite the fact the two parts of the room each have a cooling system, so it would be justified by the influence of the transit of people going in and out of the room to cause the opening and closing of the door repeated times it makes the refrigeration system work harder and makes that the NTD goes below the set point, reaching at some points less that −0.05 units. The NTD corresponds with the normalized indoor variance (NIV) where one can appreciate as well the separation by the wall and how the variance values are higher where the door on the outside is located, with an average in NIV of 0.01.

Room 11 shows a NIV always below the unit which means that the indoor variance is always lower than the outside one; this is justified by the big variability of temperatures during the Spanish summer, and the controlled temperature inside the cold rooms. The temperature in Chamber 11 was horizontally less homogeneous (having more or less an average NIV of around 0.01 in the entire room with no differentiated layers) than in Chamber 29, which has different NIV levels each with a homogeneous NIV such as 0.03, 0.02 and 0.015, which can be justified by the existence of an intermediate wall that divides the refrigerated chamber into two zones, while Chamber 29 is less homogeneous vertically having the perfectly differentiated layers mentioned above, which is justified because the Chamber 29 is only provided with one cooling unit for the same space than room 11 which is provided with two units, one per each part of the room.

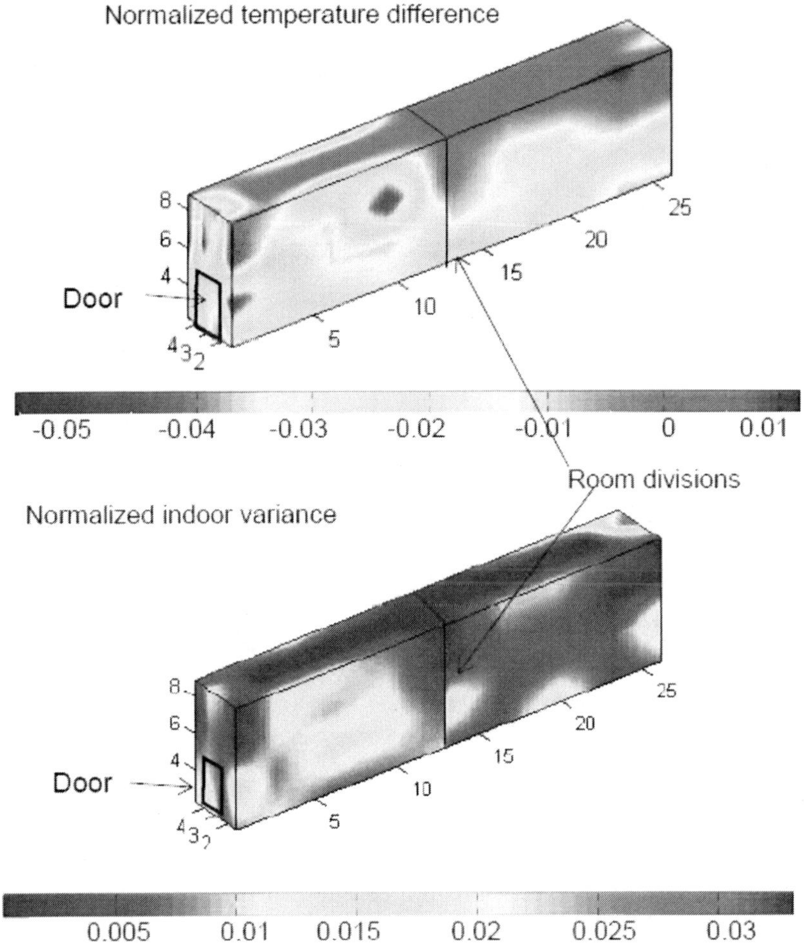


**Figure 3.** 3D plot of Normalized Temperature Difference and Indoor Variance in Chamber 11.

Room 29 as is said (Figure 4), is not provided by a separation wall, which makes the air flow along the space without interruptions, but with the characteristic that it has to cool down a space double the size of Chamber 11. NTD gets closer to the set point value close to the ceiling of the room than in the rest of it, reaching in some spots almost 0, but always below the set value, while in the rest of the room it has an average NTD of −0.13.

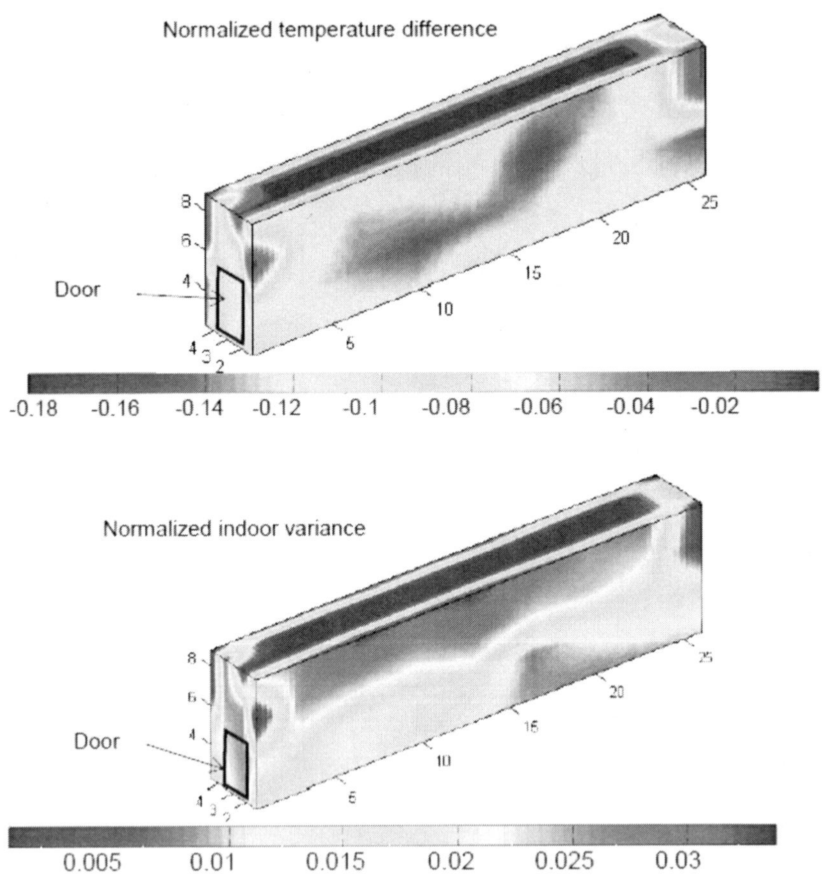


**Figure 4.** 3D plot of Normalized Temperature Difference and Indoor Variance in Chamber 29.

## 3.3. Psychrometry

The absolute humidity of the air inside the chambers was calculated, based on the ASAE standard D271.2 (1979) [20], using the data recorded during experiments. Psychrometric charts for two chambers (numbers 29 and 40) are included inFigure 5 and Figure 6, which illustrate the evolution of air absolute humidity (H, kg of water/kg of dry air) related to the T (°C). The blue cloud corresponds to the mote inside the chamber while the red one refers to the mote located outside (pre-chamber). Door openings created an increase in T (°C) and H (kg of water/kg of dry air), which then returns to normal again once the door is closed. During the rest of the time, it is also possible to detect the interaction between air properties and the product; with the cycles of cooling, variations in the absolute humidity can be estimated: condensation over the products (as loss of absolute humidity), or water evaporation (as an increase in absolute air humidity). The diagram corresponding to room 11 is similar to that of room 29, which is why is not presented in this article.

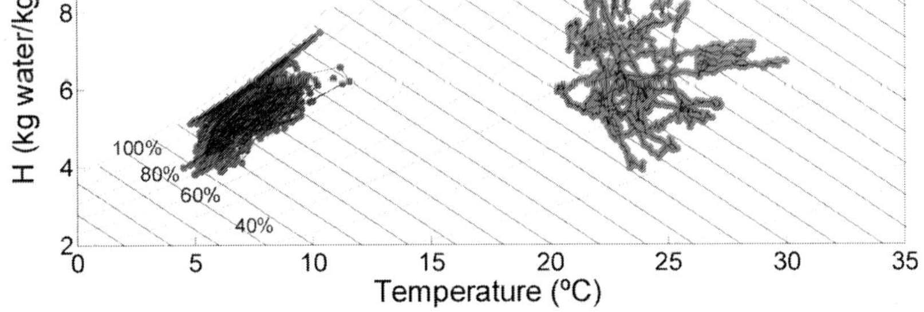


**Figure 5.** Psychrometric chart in Chamber 29.

In Figure 5, corresponding to Chamber 29, an important difference of temperatures between inside and outside the room is appreciated, as the blue cloud in many occasions reaches a RH of 100%, water saturation in the air, and this cloud of dots is always higher in RH lines than the cloud corresponding to the outside. One can appreciate differentiated paths moving up and down the RH lines, corresponding to a vapor condensation and increase of the temperature due to the door opening, when the air with different enthalpies is mixing. Both the inside node and the outside node present very similar lower H (equal to 4 kg water/kg dry air), while the upper H reaches more than 12 kg water/kg dry air in the outside node, and inside does not go beyond 8 kg water/kg dry air. There is no exchange of air flow at any moment; the air masses are clearly separated. The cloud is more compact in the interior of the room than in the exterior and it remains around 5 to 10 °C, while outside it has a wider temperature interval (from 20 to 30 °C) and with the same absolute humidity, it stays below in RH.

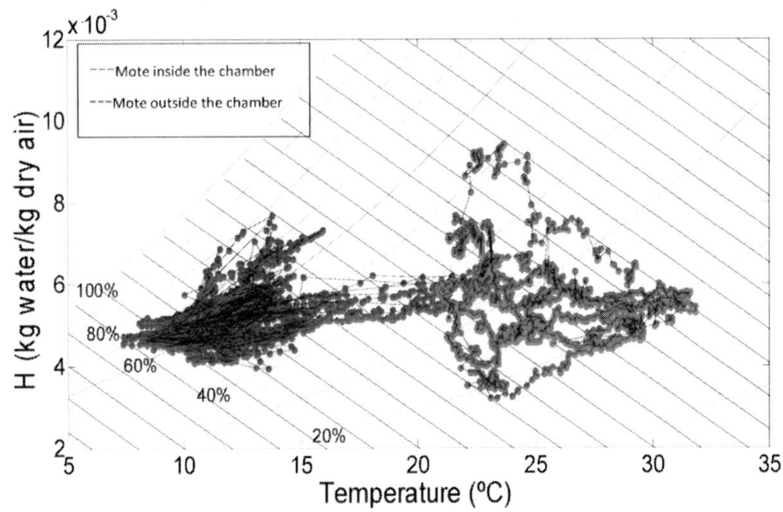


**Figure 6.** Psychrometric chart in Chamber 40.

In this case (Figure 6) the air inside has an H oscillating in a smaller range (between 4 and 8 kg water/kg dry air) than in the outside (between 3 and 10 kg water/kg dry air), and a much wider range of temperatures than the other two chambers (Figure 5) because the room was open during long periods of time; it even reaches 23 °C, which is a regular temperature registered in the node outside. The blue cloud (inside) shows horizontal movements corresponding to an increase in temperature due to the door opening. The outside cloud contains wide cycles that correspond to an opening and closing of the door and/or changes between day and night, since the pre chamber area is directly influenced by the heat of the sun. In contrast to Chamber 29, it is appreciated how the temperatures are equalized in the middle of the image due to the fact the door stays open for a long time, which is causing the mixing of the air and heat exchange, and a temperature and humidity gradient is also appreciated inside the chamber, in some occasions the RH reaches 80% whereas outside it never goes above 55% despite the door usually being open.

The use of a psychrometric equation provides very valuable information regarding air fluxes and water transport phenomena from the product to the air. Using two sensors of just one single mote it is possible to obtain such information in real time, and one can conclude by observing Figure 4 and Figure 5, both from Chamber 29, that the data obtained are complementary, since it is not possible to deduce that there is no exchange of air flow at the door by observing Figure 4alone. In addition, the temperature registered by the inside node near to the door is not representative of the entire room; therefore the RFID tags are completing the data pool for the subsequent analysis. The continuous data flow coming from the nodes can be accessed at any moment; the nodes are located in critical spots (like the doors). This information is an indicator of the surrounding area, which can be studied in detail thanks to the data obtained from the temperature tags.

# 4. CONCLUSIONS

In this paper a practical demonstration in a real scenario of the complementarity of RFID semi passive tags and the WSN is shown. The performance of the RFID tags and the WSN nodes was positive overall, the data losses was acceptable, small enough to allow one to analyze the parameters in order to create temperature distribution maps and psychrometric diagrams, but this set up is only possible in a testing phase. Since such a massive quantity of RFID tags is not possible in this scenario in the way those were positioned, it means an *ad hoc* version of this system should be developed for each case. The relative low cost of the RFID loggers allows dense implementations that provide accurate information about temperature gradients inside the chambers. The amount of data obtained from all the tags is manageable with a proper program such as devoted programs developed in Matlab. In the other hand extracting the data from a big number of these units in the cold rooms that are subject of study was a difficult task to accomplish, due to the location and the main limitations of these loggers which are their reading range and sensing capabilities.

Some of the loggers could be substituted by certain temperature estimation methods, but although these techniques are demonstrated to be reliable, other loggers located in critical spots should not be removed since there are changing conditions that cannot be predicted, such as the movement of the pallets inside the room.

Further experiments can provide a closer implementation of the two technologies, by developing a complex system comprising WSN nodes and RFID units. The benefits of this implementation are shown in this work, on the one hand WSN presents a long reading range thanks to the point to point (PtP) data transmission and a constant information flow, while in the other hand RFID lacks a PtP chain and also the tags used in the present experiments do not send information unless requested by the reader. The high price of the sensor nodes makes a dense node mesh difficult to afford, but this weakness could be resolved by using cheaper tags, and also the RFID tags are more manageable and robust due to their encapsulation.

After a detailed analysis of the data gathered in these experiments, and the experience achieved using the devices that are provided with the sensors used, the authors would like to recommend some improvements and changes to be made in the construction of the sensors. The reading range should be addressed; long range reading fixed readers or handhelds should be implemented and enhanced so communication of large amounts of data can be done with commercial RFID temperature tags. Also the range of sensing can be improved, some prototypes like the named in the introduction of this document could be turned in commercial units, as shown in this article, the use of pressure and humidity is a useful tool to see the air exchange in critical points such as the door of the cold room. As demonstrated in the experiment, WSN nodes can provide information that can be essential to detect breaks in the cold chain, nonetheless WSN nodes lack robustness, these devices are always weak and the electronics are very exposed to any possible damage, so manufacturers should work on motes able to support friction, impacts and adverse climatological conditions. Maybe some kind of hybrid unit creating a RFID sensor network, where the lightness, robustness and price from RFID is combined with the extended reading range PtP and diversity of sensors characteristic from the WSN units.

In the other hand, the authors want to provide some chamber design rules and management protocols that can save energy consumption costs in the cold room and help to extend the freshness of the cargo. As seen in the Chamber 11 the back part of it is closer to the set point than the front which is connected with the pre-chamber through a door, therefore, the most sensitive goods should be placed in the back part. In Chamber 40, a big waste of energy was noted because the outdoor and indoor air is being mixed constantly, hence the stored vegetables should be placed in the back part, and the processing tables moved to the front part. Finally an influencing factor that affects all the rooms is the opening and closing of the door, so a cooled pre-chamber should be studied; the actual pre-chamber is a corridor connected with the exterior with no isolation, and this has a certain negative effect on the temperatures when there is transit of the staff.

## ACKNOWLEDGMENTS

Thanks to SMART-QC research program (Ref.: GL2008-05267-C03-03/ALI) supported by the Ministry of Science and Innovation and TAGRALIA project funded by the Comunidad de Madrid (Ref.: S-0505/AGR-0187). The authors would also like to thank Antonio Rabasco and Jorge Sanchez for their help during the experiments.

## AUTHOR CONTRIBUTIONS

In this article, Ricardo Badia contributed with the experimental work, analyzed the data, wrote the mayor part of the article and took charge of modifying the article once written. Luis Ruiz designed the experiments and supervised them, also wrote part of the article and made the tables. Javier Garcia created the figures and analyzed the data. Jose I. Robla supervised the article and contributed to write it, as well as provide part of the valuable materials.

# REFERENCES


1. Ruiz-Garcia, L.; Lunadei, L. The role of RFID in agriculture: Applications, limitations and challenges. *Comput. Electron. Agric.* **2011**, *79*, 42–50.

2. Gwanpuaa, S.G.; Verbovena, P.; Leducqb, D.; Brownc, T.; Verlindend, B.E.; Bekelea, E.; Aregawia, W.; Evansc, J.; Fosterc, A.; Duretb, S.; *et al.* The FRISBEE tool, a software for optimising the trade-off between food quality, energy use, and global warming impact of cold chains. *J. Food Eng.* **2015**, *148*, 2–12.

3. Gustavsson, J.; Cederberg, C.; Sonesson, U.; Otterdijk, R.; Meybeck, A. Global food losses and food waste: Extent, causes and prevention. In *Interpack 2011*; Food and Agriculture Organization of the United Nations: Rome, Italy, 2011.

4. Kim, W.R.; Aung, M.M.; Chang, Y.S.; Makatsoris, C. Freshness Gauge based cold storage management: A method for adjusting temperature and humidity levels for food quality. *Food Control* **2015**, *47*, 510–519.

5. Jedermann, R.; Nicometo, M.; Uysal, I.; Opara, L.; Janssen, S.; Lang, W. *Intelligent food logistics*; The Royal Society: London, UK, 2015.

6. Sánchez López, T.; Daeyoung, K. Wireless Sensor Networks and RFID Integration for Context Aware Services. In*White Paper Series 2008*; AUTOIDLABS-WP-SWNET-026; Daejeon, Korea, 2008. [Google Scholar]

7. UN/FAO 2013. Food Wastage Footprint Impacts on Natural Resources: Summary Report. 2013. Available online: http://www.fao.org/docrep/018/i3347e/i3347e.pdf (accessed on 12 September 2014).

8. Mahajan, P.V.; Caleb, O.J.; Singh, Z.; Watkins, C.B.; Geyer, M. Postharvest treatments of fresh produce. *Philos. Trans. R. Soc. A Math. Phys. Eng. Sci.* **2014**.

9. Jedermann, R.; Nicometo, M.; Uysal, I.; Lang, W. Reducing food losses by intelligent food logistics. *Philos. Trans. R. Soc. A Math. Phys. Eng. Sci.* **2014**, *372*, 20130302.

10. Lian, F.; Zhao, W.; Yang, R.; Tang, Y.; Katiyo, W. Survival of *Salmonella enterica* in skim milk powder with different water activity and water mobility. *Food Control* **2015**, *47*, 1–6.

11. Burg, S.P. Chapter 10—Capillary Condensation in Non-Waxed Cardboard Boxes. In *Hypobaric Storage Food Industry: Advances in Application and Theory*; Academic Press: London, UK, 2014.

12. Zou, Z.; Chen, Q.; Uysal, I.; Zheng, L. Radio frequency identification enabled wireless sensing for intelligent food logistics. *Philos. Trans. R. Soc. A Math. Phys. Eng. Sci.* **2014**, *372*, 20140209.

13. Nascimento Nunes, M.C.; Nicometo, M.; Emond, J.P.; Badia-Melis, R.; Uysal, I. Improvement in fresh fruit and vegetable logistics quality: Berry logistics field studies. *Philos. Trans. R. Soc. A Math. Phys. Eng. Sci.* **2014**, *372*, 20140212.

14. Fiddes, K.; Yan, N. RFID tags for wireless electrochemical detection of volatile chemicals. *Sens. Actuators B Chem.* **2013**,*186*, 817–823

15. Badia-Melis, R.; Garcia-Hierro, J.; Ruiz-Garcia, L.; Jiménez-Ariza, T.; Robla Villalba, J.I.; Barreiro, P. Assessing the dynamic behavior of WSN motes and RFID semi-pasive tags for temperature monitoring. *Comput. Electron. Agric.***2014**, *103*, 11–16.

16. Georg, H.H.; van Jos, H.; Eelco, J.; Klaas, S. Adoption of Wireless Sensors in Supply Chains: A Process View Analysis of a Pharmaceutical Cold Chain. *J. Theor. Appl. Electron. Commer. Res.* **2013**, *8*, 138–154.

17. Ruiz-Garcia, L.; Steinberger, G.; Rothmund, M. A model and prototype implementation for tracking and tracing agricultural batch products along the food chain. *Food Control* **2010**, *21*, 112–121.

18. Qi, L.; Xu, M.; Fu, Z.; Mira, T.; Zhang, X. C2SLDS: A WSN-based perishable food shelf-life prediction and LSFO strategy decision support system in cold chain logistics. *Food Control* **2014**, *38*, 19–29.

19. Jedermann, R.; Pötsch, T.; Lloyd, C. Communication techniques and challenges for wireless food quality monitoring.*Philos. Trans. R. Soc. A Math. Phys. Eng. Sci.* **2014**, *372*, 20130304.

20. ASABE. Psychrometric Data ASAE D271.2 APR1979, R2005.

21. Uysal, I.; Emond, J.P.; Bennett, G. Tag testing methodology for RFID enabled temperature tracking and shelf life estimation. In Proceedings of the IEEE International Conference on RFID Technology and Applications, Sitges, Spain, 15–16 September 2011.

22. Jedermann, R.; Lang, W. The minimum number of sensors—Interpolation of spatial temperature profiles in chilled transports. *EWSN* **2009**, *LNCS 5432*, 232–246.

23. Badia-Melis, R.; Brecht, E.; Lowe, A.; Uysal, I. Pallet wide temperature estimation and mapping for perishable food. In Proceedings of the Poster on Annual Conference ASABE 2013, Kansas City, MO, USA, 21–24 July 2013.

24. Ruiz-Garcia, L. Development of Monitoring Applications for Refrigerated Perishable Goods Transportation. Ph.D. Thesis, Ingenieria Rural, Universidad Politécnica de Madrid, Madrid, Spain, 2008.

11

# ACCURATE MONITORING AND FAULT DETECTION IN WIND MEASURING DEVICES THROUGH WIRELESS SENSOR NETWORKS

**Komal Saifullah Khan* and Muhammad Tariq**

Department of Electrical Engineering, National University of Computer & Emerging Sciences (NUCES), Peshawar Campus, Peshawar 25000, Pakistan

## ABSTRACT

Many wind energy projects report poor performance as low as 60% of the predicted performance. The reason for this is poor resource assessment and the use of new untested technologies and systems in remote locations. Predictions about the potential of an area for wind energy projects (through simulated models) may vary from the actual potential of the area. Hence, introducing accurate site assessment techniques will lead to accurate predictions of energy production from a particular area. We solve this problem by installing a Wireless Sensor Network (WSN) to periodically analyze the data from anemometers installed in that area. After comparative analysis of the acquired data, the anemometers transmit their readings through a WSN to the sink node for analysis. The sink node uses an iterative algorithm which sequentially detects any faulty anemometer and passes the details of the fault to the central system or main station. We apply the proposed technique in simulation as well as in practical implementation and study its accuracy by comparing the simulation results with experimental results to analyze the variation in the results obtained from both simulation model and implemented model. Simulation results show that the algorithm indicates faulty anemometers with high accuracy and low false alarm rate when as many as 25% of the anemometers become faulty. Experimental analysis shows that anemometers incorporating this solution are better assessed and performance level of implemented projects is increased above 86% of the simulated models.

## KEYWORDS

wind speed monitoring; error detection; wireless sensor networks; site assessment; cup anemometers

## 1. INTRODUCTION

Over the past few years energy production through wind has increased dramatically because of the wind's potential economic, environmental and security benefits. Wind energy resources provide cheaper and economical electricity as compared to other energy resources. In the US wind farms are generating electricity at a rate less than 5 cents per kilowatt [1]. Due to the fast increase in wind technology deployment, developing projects require much better and accurate monitoring along with supervising techniques. Very accurate wind speed measurements are very important to the wind energy community. According to [1], many installed projects report poor performance due to poor resource assessment and poor fault prediction techniques. Errors in wind speed measurement are one of the major reasons for causing uncertainties in power generation performance of wind turbines. Accurate monitoring of parameters (like wind speed, direction, overspeeding, number of rotations, torque, turbulence intensity and temperature) is therefore a major contribution to the wind energy community.

Wind measuring devices like anemometers are used for evaluating a potential site for wind energy projects. Regular maintenance and accurate monitoring of anemometers, installed for site assessment, is one of the main challenges for wind energy deployment companies because of the high costs required for it. Errors occurring in the readings taken by these devices are usually checked manually by sending labor to the area where these devices are installed. No proper monitoring techniques for these devices are present in order to get accurate measurements by these devices which in turn results in inaccurate assessment of the site. This badly affects the projects that are to be implemented in that area as the projects when implemented perform far less than expected. In our analysis and experimentation, a WSN will play a vital role in monitoring and supervising wind resources in remote areas and reporting the data back to the central control station. A number of different locations can be monitored through one station. This network will minimize the cost of cables and the labor that will be needed to manually check the sites and other costly accessories.

Cup anemometers are accepted by international standards for power performance measurements and are used widely in wind energy applications [2]. Deployments in remote areas require regular engagement and more oversight than conventional wind energy projects [3]. According to National Renewable Energy Laboratory (NREL) in the US, many wind energy projects report results as low as 60% of the predicted performance [1]. The reason for this is poor resource assessment and the use of new untested technologies and systems in remote locations. Hence, better assessment and fault detection techniques are required to achieve the desired performance level. For a balanced and economical system, it is important to reduce operational costs of the project. Condition-based maintenance systems will help in achieving accurate and improved performance in energy projects to be implemented and will decrease the operational costs. All wind turbines installed worldwide by the end of 2008 are generating 260 Terawatt hour (TWh) per annum, corresponding to more than 15% of the global energy consumption [4]. It is clear that in the mid- to long-term, wind energy investments will rather be strengthened due to their low risk character and additional economic benefits. The contributions of this paper are as follows:

- We are the first to introduce a WSN for accurate monitoring and fault detection in wind measuring devices. We have used our own developed sensor nodes instead of traditional sensor nodes such as MICAZ, TelosB, or IRIS, which does not suit our experiments due to limited radio range and sensing capabilities.

- This paper proposes a technique for accurate site assessment. Proper assessment of a site will lead to accurate estimations of the energy production from future wind energy projects to be implemented.

To the best of our knowledge, no work has been done yet on implementing a WSN for site assessment purposes. The remainder of the paper is organized as follows: Section 2 explains the related work in detail. Section 3 explains the design description and Section 4 explains the proposed algorithm. Section 5 demonstrates the simulation and experimental results. Finally we conclude the paper in Section 6.

# 2. RELATED WORK

Some significant work that has been done on monitoring fault issues in anemometers is summarized below.

## 2.1. Fault Detection Techniques in Sensors and in Wind Measuring Devices

Good quality work has been done on detecting faults in sensors. In [5], failed nodes were traced at the base station, assuming that all sensor measurements will be directed to the base station through a routing tree. In this work, the base station has a global view of the network topology and it can identify failed nodes through route update messages. In recent works for fault detection [6–8], sensors are required to access a binary decision procedure by comparing their readings with a predetermined threshold. This binary decision is called a 0/1 decision predicate with 0 indicating a normal status and 1 indicating a faulty status.

Some research work has also been done in identifying faults in wind measuring devices. A technique for data analysis was discussed in [9] to identify faults in a wind turbine installed in a wind farm. The data analysis for this purpose was done through automated Supervisory Control and Data Acquisition (SCADA). In [10], an algorithm for fault detection in wind turbines was discussed. Condition monitoring techniques were introduced in order to detect faults in different components of the turbine. Sensors were also used for environmental monitoring applications. Also, for environmental conditions monitoring, a WSN deployment was carried out to monitor environmental conditions [11]. This technique relied on installing a low cost station in the alpines in Switzerland which was a better technique for environmental monitoring as compared to high cost stations installed for this purpose. In [12], an anemometer-based method for remote area wind energy harvesting was presented. The motion of the anemometer shaft was utilized to turn an alternator which in turn would be useful for harvesting energy.

In the past a lot of work has been done for developing wind measuring devices that will accurately measure wind speed. In 1990, a one dimensional, ionic flow anemometer was developed in Japan using a high voltage pulse [13]. The air molecules were ionized and the flow time detected for the ionized molecules was used for wind speed calculation. But by using this technique only single direction measurements could be made and the right direction was obtained by trial and error method. Another one dimensional, fixed electric field measurement technique was applied in 1995 [14]. Air molecules were ionized using high voltage between polar plates in a fixed distance. In 2009, a mobile optical coordinate measuring technology was used to acquire offshore wind turbine data and behaviors [15].

## 2.2. Site Assessment Techniques

Site assessment for wind energy projects can also be done with the use of wind maps. In 2011, a Geographic Information System (GIS)-based wind farm site assessment technique was used. Using this method, large geographic areas could be assessed and the evaluation outcome could be displayed in suitable maps [16]. Wind atlases are also used for site assessment. In [17], a wind resource atlas was prepared where wind observations were recorded from meteorological stations around that area. These recorded observations were used to define the statistical details of the wind resource in that particular area. However, as these maps short fall of giving accurate results, these are useful only for initial selection of sites. Although improvements have been made to obtain accurate results by these maps, it is still improbable that these maps will end the importance and need for on-site measurements for site selection. However, this technique can still help in speeding up the process for proper site selection. Apart from wind maps, Laser Imaging Detection and Ranging (LIDAR) and Sonic Detection and Ranging (SODAR) are also used for site assessment. SODAR and LIDAR are basically remote sensing (RS) devices that use sound and light, respectively. These technologies are ground-based and are capable of working for quite long distances which make them easy to use for analyzing and evaluating a site. In September 2013, wind speed data was collected at Buena Vista Wind Farms in California using a LIDAR instrument (Zephir 300) [18]. However, the LIDAR required more power than expected and measured high wind speeds at night. The reason for this was that the LIDAR was not placed on truly flat land. Hence, these ground based RS devices show variable and unpredictable measurements, especially in hilly terrains. Improvement of these devices is very important in order to increase the confidence in using these ground based techniques [19]. Up till now, wind site assessment process has counted on cup anemometers and approved it as a standard for this purpose [2].

# 3. DESIGN DESCRIPTION

Before explaining the design details, we first give background information about the cup anemometers that will be used. Then we explain the comparison of anemometer with a potential RS technology, SODAR. After that we give the details of proposed design.

## 3.1. Background

An anemometer is a device that is used for measuring wind speed in any particular area. Typically, anemometers are deployed in an area for assessing the potential of a site for future wind energy projects. By using anemometers, we can analyze the following parameters:

### 3.1.1. Wind Speed and Direction Analysis

For cup anemometer, we consider a three dimensional wind speed vector [20]. The 3 dimensional wind speed vector is given as:

$$\vec{V} = (x, y, z) \tag{1}$$

where:

$x$ = *Longitudinal Component*

$y$ = *Traversal Component*

$z$ = *Vertical Component*

Hence:

$$\left|\vec{V}\right| = \int_t \sqrt{x^2 + y^2 + z^2} \tag{2}$$

In most cases, the vertical component is excluded which makes the equation as:

$$\left|\vec{V}\right| = \int_t \sqrt{x^2 + y^2} \tag{3}$$

The difference between these two wind speed values in Equations (2) and (3) is based on turbulence intensity (Ti). For 15% turbulence, the difference is 0.5% and for 30% turbulence the difference is 1% [20]. The difference increases with high turbulence. In our analysis, we use the three dimensional wind speed vector.

We will focus on the three dimensional component of wind for accuracy. So, according to [21], the instantaneous wind speed is decomposed into its components. We assume that mean wind speed is horizontal:

$$\begin{Bmatrix} x \\ y \\ z \end{Bmatrix} = \begin{Bmatrix} V_f + x \\ y \\ z \end{Bmatrix} \tag{4}$$

where $V_f$ represents the fluctuations in wind speed. We assume that these fluctuations are small as compared to mean wind speed. The direction, θ for the three dimensional wind speed will become:

$$\theta = \tan^{-1} \frac{z}{\sqrt{x^2 + y^2}} \tag{5}$$

### 3.1.2. Effect of Turbulence

Turbulence intensity is dependent on mean wind speed and standard deviation. Lower the wind speed, higher will be the turbulence intensity:

$$T_i = \frac{\sigma}{\bar{V}} \tag{6}$$

where, σ is the standard deviation of the wind speed and $\bar{V}$ is the normalized wind speed.

For continuous gusts [22], we analyze the three dimensional turbulence:

$$T_{i,x} = \sqrt{\frac{\Phi_x(\xi)\pi}{2\ell_x}[1+(\ell_x\xi)^2]^2} \tag{7}$$

$$T_{i,y} = \sqrt{\frac{\Phi(\xi)\pi}{2\ell_y}\frac{[1+4(\ell_y\xi)^2]^2}{1+12(\ell_y\xi)^2}} \tag{8}$$

$$T_{i,z} = \sqrt{\frac{\Phi_z(\xi)\pi}{2\ell_z}\frac{[1+4(\ell_z\xi)^2]^2}{1+12(\ell_z\xi)^2}} \tag{9}$$

where ξ is the spatial frequency, Φ is the power spectral density and ℓ is the length scale.

### 3.1.3. Number of Rotations Analysis

The rotation rate for the anemometer [22] plays an important role in detecting faults in the nodes:

$$\dot{N}_r = F(N_r, \sqrt{x^2 + y^2}, z) \tag{10}$$

At steady state and constant rotation rate:

$$F(N, V, 0) = 0 \tag{11}$$

We obtain:

$$N_r = N_r(V) \tag{12}$$

The relation between wind speed U and number of rotations, $N_r$ will become:

$$V \propto N_r \tag{13}$$

We observe that for a specific wind speed:

$$N_{rx} = N_{r,rec} \tag{14}$$

where:

$N_{rx}$ = Number of rotations for node $x$

$N_{r,rec}$ = Recorded standard value for number of rotations at the given wind speed

### 3.1.4. Maximum Overspeeding Level

Overspeeding is another major reason for fault in anemometers.The maximum overspeeding level [20] for cup anemometer is dependent on the drag ratio and turbulence intensity:

$$O_{s,\infty} = \frac{2\sqrt{h} - \sqrt{4h - (1-h)^2 (\frac{\Delta V}{V_0})^2}}{(1-h)^2} \tag{15}$$

where $h$ is the drag ratio:

$$h = \frac{C_{DL}}{C_{DH}}$$

$$\frac{\Delta V}{V_0} = T_i$$

The equation becomes:

$$O_{s,\infty} = \frac{2h - \sqrt{4h - (1-h)^2 T_i^2}}{(1-h)^2} \tag{16}$$

By using second order Taylor series expansion around $T_i = 0$, we get the simple relation:

$$O_{s,\infty} = \frac{(1+h)^2}{4h} T_i^2 = \frac{T_i^2}{(1-\lambda_0)(1+\lambda_0)} \tag{17}$$
$$O_{s,\infty} = F_0 \cdot T_i^2$$

The maximum overspeeding ratio, $F_0 = 1.15$ at a speed ratio of 0.3, the normal speed ratio of cup anemometers [20].

Also, it is clear from above Equation (17) that:

$$O_{s,\infty} \alpha T_i^2 \tag{18}$$

which clearly shows the relation between wind speed and turbulence intensity.

## 3.2. Comparison of Anemometer with Other Technologies

Before explaining our experimental analysis, we discuss the comparison of SODAR, an RS technology used for site assessment with cup anemometers. Ground-based RS techniques are also used for site evaluation nowadays. These devices can be validated by comparing the winds estimated by these devices with those recorded at the same heights by cup anemometers. Mathematical analysis was done to get Figure 1. We used our own data for anemometer while already present data for SODAR [19]. The slope is plotted using equation (19). Results show that these ground based RS devices underperform as compared to cup anemometers. In hilly terrains or areas where hindrance is sufficiently great, these devices generally record lower wind speeds [19] and are therefore suitable for use in flat terrains. The variation in the wind speed calculations performed by a node (cup anemometer) and a SODAR is shown by the regression slope in Figure 1. From the graph, it can be seen that at lower wind speeds the variation is more. However, at higher wind speeds, there appears to be a decrease in the regression slope. The range of wind speeds is according to the Beaufort scale commonly used for weather forecast by meteorological stations.

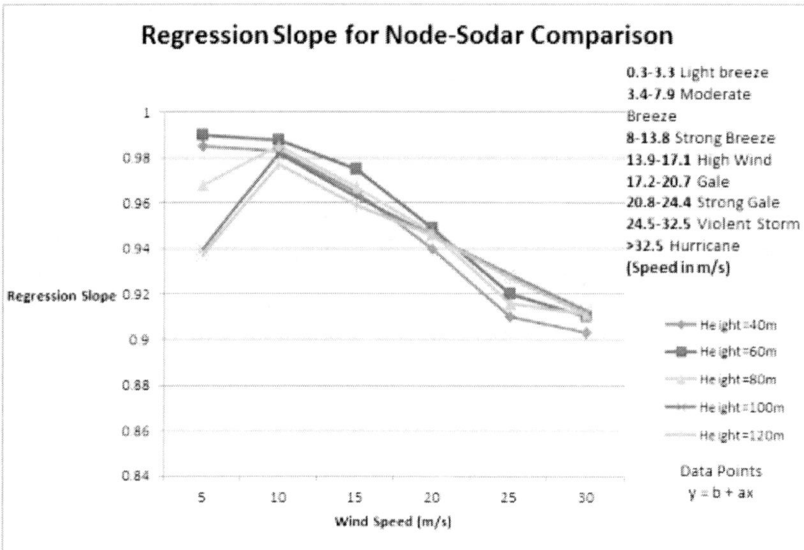


**Figure 1.** Comparison of site assessment technologies.

Also in Figure 1:

$$\text{slope, b} = \frac{N\sum XY - (\sum X)(\sum Y)}{(N\sum X^2 - (\sum X)^2)} \tag{19}$$

and:

$$\text{intercept, a} = \frac{\sum Y - b(\sum X))}{N} \tag{20}$$

## 3.3. Design Details

The design focuses on a number of anemometers deployed in a remote location which we want to assess as a potential site for deployment of wind energy project and where manual monitoring of the anemometers and fault detection becomes hard to achieve and costly. Each anemometer is equipped with a 3DR radio which operates at 433 MHz and has a range of one mile. Anemometers are therefore placed one mile apart from each other. However a large scale deployment can be done where anemometers at same heights can be placed in same clusters and can be equipped with radios that cover larger distances so that larger sites can be easily covered. For site assessment, cup anemometers are deployed in a remote location and monitoring of these anemometers is performed using a WSN. The anemometer is equipped with a 3DR as a transceiver, a memory card as a storage unit, an Arduino Uno Microcontroller [23] as a processing unit and a photo coupler as a sensing unit. The node deployment is deterministic based on same heights criteria. Nodes at same height communicate with each other for data sharing. In Figure 2, nodes placed at same height are represented by same pattern circles. Base station is deployed in the center of the assessment site while the radio range for each transceiver is 1·km. The number of nodes will increase if a vast field is to be assessed. Vast fields will require more nodes to be deployed.

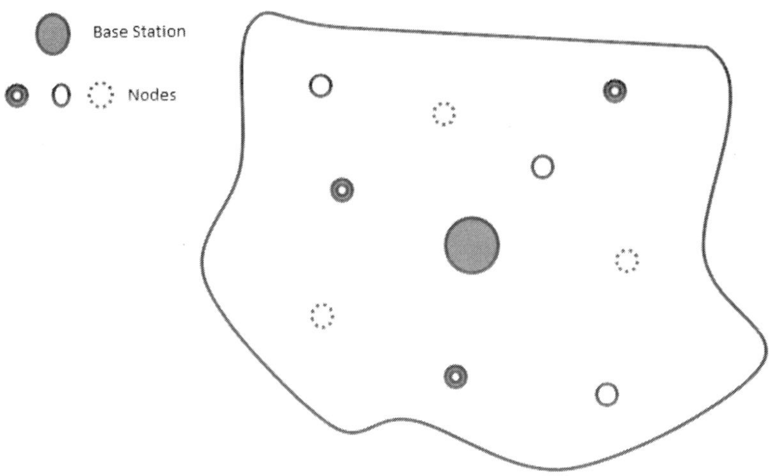


**Figure 2.** A general overview of node placement.

# 4. PROPOSED ALGORITHM

In this section, we will explain our proposed algorithm for detecting the faulty nodes in the network.

The algorithm is divided into three parts: Algorithm 1 is for developing a monitoring system to monitor the performance of the nodes,

Algorithm 2 works for detecting the faulty node and Algorithm 3 is for detecting the error due to which the fault occurred. For this purpose the task has been divided into five parts:

- Data acquisition by the nodes
- Preprocessing of the data between nodes located at same heights
- Data analysis and transmission to the sink node
- Faulty node detection at the sink node
- Data transmission to the main station via GSM

The nodes located at same height are referred to as neighbors. The nodes gather data such as wind speed, direction, number of rotations, turbulence intensity and temperature. In the second step, the node encapsulates the data with its node ID and forwards the data to its neighboring node. A node regularly shares data three times in every twenty four hours *i.e.*, nodes share data with each other once in every eight hours. However a node (cluster head) randomly decides when to share data during those eight hours. All other nodes in the cluster collect data at that time and then comparisons are made. A random variable is used in the algorithm to choose any random time during those eight hours for data sharing. The neighboring nodes share this data with each other and comparisons are made. The comparison technique is explained in Algorithm 1. If the difference between the data compared is less than a threshold percentage of 2% then the node's status is considered to be ok else the error is considered to be significant and the node transmits these results to the sink node for further analysis. Same heights criterion discussed above (Section 3.3) for getting same measurements is quite favorable in conditions like cold fronts where cold air moves rapidly to replace warmer air. In such situations, where vertical motion of air takes place, all the nodes in the same network will experience same changes in readings. Due to this the probability of false detections will be low because the algorithm relies on comparison of the readings taken by all the nodes at same height in the network.

The algorithm is divided into three parts.

- Monitoring
- Faulty Node Detection

- Fault Detection

Readings from faulty sensor nodes are geographically independent. But readings from sensor nodes in close proximity are correlated [8]. We assume that faults in the nodes do not occur regularly rather it is an event that occurs rarely. Our monitoring system is less complicated so that the nodes have to do little processing and hence save energy which will further increase the network lifetime. As the site assessment is usually expected to last for at least a year so a prolonged network lifetime would be in favor of the assessment process. In the opposite case, replacing energy drained sensors every now and then might increase the cost of the assessment project. Once a faulty node reading is detected, the results are sent to the sink node to accurately detect the error in the readings. Once a faulty node is detected, professionals can be sent to either fix the device. If the device cannot be fixed then it can either be replaced or removed from the network. We are directly working on the data collected to detect the errors in the anemometers being installed on site because the data collection process by these nodes for site assessment is carried out for months. Using this data collected, we can also find out about the node that is not giving accurate readings and can further work on correcting that node. The reason for this is that we will need that node to collect data in the coming months.

**Algorithm 1**

/* Monitoring System

/* This algorithm predicts faults and sends findings to sink node

**Input**: $V$: Wind speed

**Input**: $\theta$: Angle of wind incident upon the cup rotor

**Input**: $N_r$: Number of rotations

**Input**: $T_i$: Turbulence Intensity

**Input**: $O_s$: Overspeeding

**Input**: $\tau$: Torque

**Input**: $t$: Temperature

**Output**: monitored fault sent to sink node

/* A uniform random variable x is used to share data between two nodes at random intervals

/* At some random time:

Node id ← Data ($V$, $\theta$, $N_r$, $T_i$, $O_s$, $\tau$, $t$)

    Send to Node 2

/* CP = comparison results of node x and (x+1) data

    At Node 2:

    if (CP < Threshold)

    Node status = OK

    end

else

    if (CP > Threshold)

    Send results to sink node

    end

For accurate analysis at the sink node we consider an $a \times a$ squared field where N nodes are uniformly deployed. A sink node is that node in a WSN where data from all other nodes is collected. Sink node will process the data and will send it to the main station. Each node $n_i$ records measurements at regular intervals. Let $m_i$ denote the reading of the sensor node $n_i$and $m_i = \{V, \theta, N_r, T_i, O_s, \tau, t\}$. Now let N $(n_i)$ denote a bounded closed set in the region that contains the node $n_i$ and additional $k$ sensors $n_1, n_2, ..., n_k$. Let $m_1, m_2, ..., m_k$, denote the measurements taken by the nodes $n_1, n_2, ..., n_k$ respectively. At the sink node comparison will be done between the reading of the node $n_i$ and the center of $m_1, m_2, ..., m_k$. The difference is:

$$d_i = m_i - med_m \tag{21}$$

where $med_m$ denotes the median of the measurements taken. It should be clear that $d_i$ will contain not a single value, but rather multiple values as a result of the comparison of all the parameters discussed above. Similarly, we find the difference for all the sensor nodes in $N(n_i)$ using Equation (22) to yield $D = \{d_1, d_2, ..., d_i ..., d_k\}$. Let $\hat{\mu}$ and $\hat{\sigma}$ denote the sample mean and sample standard deviation of the set $D$, *i.e.,*:

$$\mu = \frac{1}{n}\sum_{i=1}^{n} d_i \tag{22}$$

$$\hat{\sigma} = \sqrt{\frac{1}{n-1}\sum_{i=1}^{n}(d_i - \hat{\mu})^2} \tag{23}$$

Standardize the dataset $D$ to eliminate the errors and to obtain $\{s_1, s_2, ..., s_n\}$, where:

$$s_1 = \frac{d_1 - \hat{\mu}}{\hat{\sigma}}, \ldots\ldots s_i = \frac{d_i - \hat{\mu}}{\hat{\sigma}}, \ldots s_n = \frac{d_n - \hat{\mu}}{\hat{\sigma}} \tag{24}$$

Now decision will be made based on this data.

If $|s_i| \geq$ *Thershold*, treat $n_i$ as faulty node and assign $s_i$ to $F_d$ where $F_d$ is the set of sensors that are claimed as faulty by the above procedure.

**Algorithm 2**

/* This algorithm explains faulty node detection technique to be used at sink node.

**Input**: Data ($V$, θ, $N_r$, $T_i$, $O_s$, τ, $t$) from nodes

**Output**: Faulty node detected

1 For each node $n_i$, construct N ($n_i$)

2 Compute $d_i$ for node $n_i$ using $N(n_i)$ and Equation (21)

3 Compute $s_i$ for node $n_i$

4 If $s_i \geq$ *Thershold*, treat $n_i$ as a faulty node

**Algorithm 3**

/* Fault Detection

**Input**: Faulty node data

**Output**: Cause of fault

for ($V_r = V_{r,rec}$)
if ($N_{rx} > N_{r,rec}$)
Fault = under speeding
/* Check temperature and turbulence effects
if (temperature < threshold value)
Fault = Low temperature
else
Fault = Turbulence effect
else
if ($N_{rx} < N_{r,rec}$)
Fault = Over speeding
Send results to main station
end

# 5. SIMULATION AND EXPERIMENTAL WORK

In this section, we first focus on analyzing the fault detection algorithm by simulating it using Aeolus SimWindFarm Model [24] and then implement our algorithm on our experimental setup. We then compare the simulated model with the actual experimental model to check the variation in results of simulated and implemented model and check the fault detection accuracy of the WSN. Our goal is to achieve less variation in results of experimentation and simulation results.

## 5.1. Simulation and Results

Our evaluation and results for now are based on simulation. Simulations using Matlab® have been performed for obtaining the results. As mentioned before, the wind simulation and modelling is done using the Aeolus SimWindFarm toolbox [24] which is a deterministic toolbox. We implemented our algorithm on this wind farm model. Using this model we have built a network to evaluate their performance. Each simulation run follows some scenario. Each scenario is based on wind speed ranges. Wind speed ranges are given in Figure 1. All the parameters (as discussed in Section 3.1) are analyzed at these different wind speeds. The accurate analysis of these parameters will show the accuracy level of the anemometers. Ten runs are performed at each wind speed range. For example, ten runs are performed for light breeze; ten runs are performed for moderate breeze and so on. The parameters used for simulation are given in Table 1.

**Table 1.** Simulation parameters and their values.

| Parameter | Value |
|---|---|
| Field size | 10 km × 10 km |
| Number of nodes deployed | 50 |
| Topology | Star topology with sink node and base station at the center |
| Radio range | 1.6 km |
| Packet size | 40 Bytes (Data = 32 Bytes + Control 8 Bytes) |
| Simulation time | 400 s |

The faulty nodes are detected by the WSN to reduce the fault level in the results. To evaluate the performance of $F_d$, we calculate the detection accuracy $a(F_d)$ and the false alarm rate, $e(F_d)$:

$$a(F_d) = \left| \frac{F_d \cap F}{F} \right|, \; e(F_d) = \left| \frac{F_d - F}{N - |F|} \right| \qquad (25)$$

If $a(F_d)$ is high and $e(F_d)$ is low, Algorithm 2 has good performance. Note that when the probability of faulty node, $p < 0.25$ and density of the network *i.e.*, the number of nodes is ≥30, then according to Figure 3, the accuracy is above 86% and in Figure 4, the false alarm rate is around 1%–2%. Increasing the number of nodes can take the accuracy level up to 89%. Although an accuracy level of 86% is achieved but it is surely not enough. The more higher the accuracy level, the more accurate assessment can be done about a site. 100% accuracy will mean that the prediction done about the energy production estimate of a particular site was 100% correct. Lower this percentage, lower will be the energy production than the estimated one. From Figure 3 it is clear that increasing the number of nodes can increase the probability of accurate assessment of a site hence taking the percentage to a much higher level. When the number of nodes increases, the accuracy level also increases. As it is a low level experimentation so we expect that a more accurate assessment can be done in future by using more nodes in the experimentation.

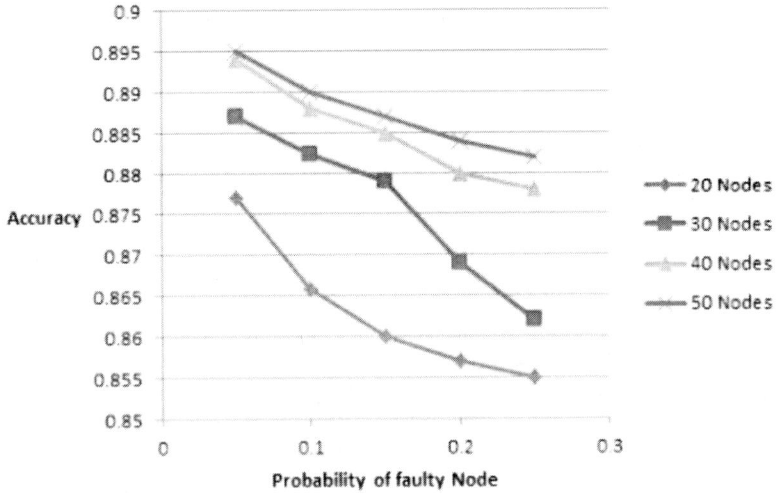


**Figure 3.** Probability of faulty node *vs.* accuracy.

The results can also be improved by recording data more than three times a day. As a node's measurement is compared with the average of the recorded values so a much better average can be achieved by recording data more than once in a particular slot. For future work, it might be interesting to check the probability of getting accurate readings with increasing the number of measurements taken in a slot.

## 5.2. Experimentation

After evaluating the network using simulations, we implemented it using hardware. The anemometers were deployed in an outdoor environment along with a WSN. Each anemometer is equipped with our own designed sensor node which comprises of a photo coupler working as a sensing unit and a 3DR radio as transceiver. This WSN constantly monitors the performance of the nodes installed and sends the results to the main station via Global System for Mobile Communication (GSM). The anemometers that will share data with each other are carefully placed at the same height because anemometers at different heights will produce different readings. Our experimental analysis is a low level implementation and is therefore based on deployment of three anemometers carefully placed at the same height. We did not implement all features. To test the effectiveness of the scheme, large scale experimentation is required, involving a lot of nodes. During testing, we implemented a network with three nodes to analyze the performance through WSN. Simulation performed with fewer nodes was compared with the experimental results to compare the performance difference accurately. Data from this implemented setup is usually observed for one year. From this recorded data the annual frequency distribution of wind speed is calculated. However, we performed on site measurements for three months starting from 3 February 2014 to 15 May 2014 and worked on the results obtained. We exercised all scenarios with different wind speeds to evaluate the accuracy of the model. For wind energy related applications, the range in which the wind speed can be measured with a very high accuracy is 4–16 m/s. Below 4 m/s threshold, the wind power is almost negligible and is not worthy of consideration for the wind turbines while above 16 m/s, the power becomes almost constant and therefore can be eliminated from consideration. We consider that three nodes implementation play an important role in analyzing a potential area for wind farms, however implementation with more nodes are highly desired.

Equation (6) shows that low wind speeds usually generate higher turbulence because of the inverse relation between mean wind speed and turbulence intensity. Hence low wind speed will lead to high turbulence in the anemometers and data collected by the anemometer might have variations. A threshold value of 2% is therefore set to detect significant errors and ignore the negligible errors in the analysis of the received data. It should be noted here that if the threshold value is set below 2% then even the minor errors will be detected as major errors and the node will be declared as faulty (false positive detection) which will increase the false detection rate. If the threshold value is greater than 2% then major errors might be missed and missed detections will occur.

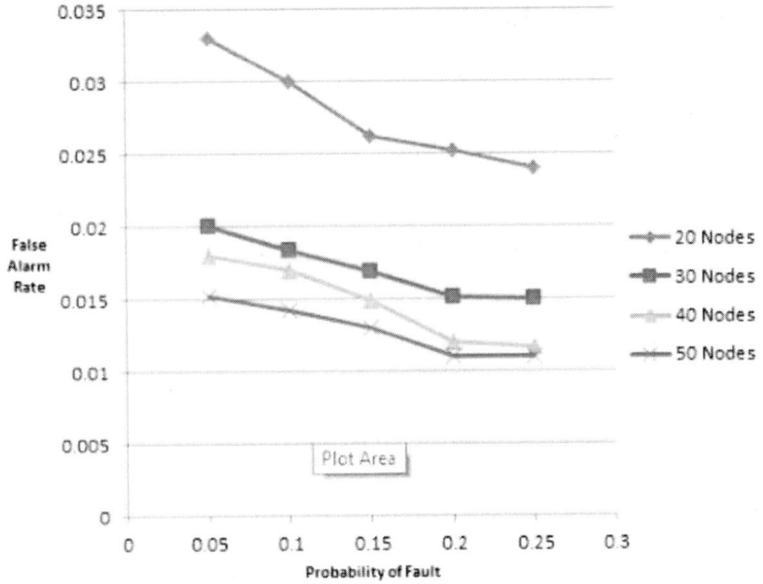


**Figure 4.** Probability of fault *vs.* false alarm rate.

A node shares data packets at regular intervals with other nodes located at the same height. The following parameters are shared between the nodes:

- Wind speed, $V$
- Direction, $\theta$
- Number of rotations, $N_r$
- Turbulence intensity, $T_i$
- Overspeeding, $O_s$
- Torque, $\tau$
- Temperature, $t$

The data shared is eventually sent to the sink node for further analysis. These parameters play an important role in detecting and correcting faults in the system. The proposed algorithm analyzes these parameters to evaluate the performance of the nodes.

A WSN is deployed to monitor the accuracy level of the anemometers and make sure that the readings taken by the anemometers are accurate. This will help in evaluating the site properly. We assume that each anemometer can compute its position through either GPS or some other techniques as mentioned in [25–27]. We use EARS routing scheme [28], a scheme which relies on radio information to route data through the network to the sink node. Cup anemometers are robust, easy to operate and are resistant to turbulence. We refer to the anemometer as a node in the remainder of the paper.

Figures 5 and 6 show the three cup anemometer used for experiment. As discussed above, each anemometer is equipped with our own designed sensor node. Figure 5 shows the photo coupler attached to the anemometer for sensing purpose. The relation between mean wind speed and turbulence intensity according to IEC standards [29] is shown in Figure 7. It should be observed from Figure 7 that at low wind speeds, turbulence intensity is high while at higher wind speeds this intensity decreases.

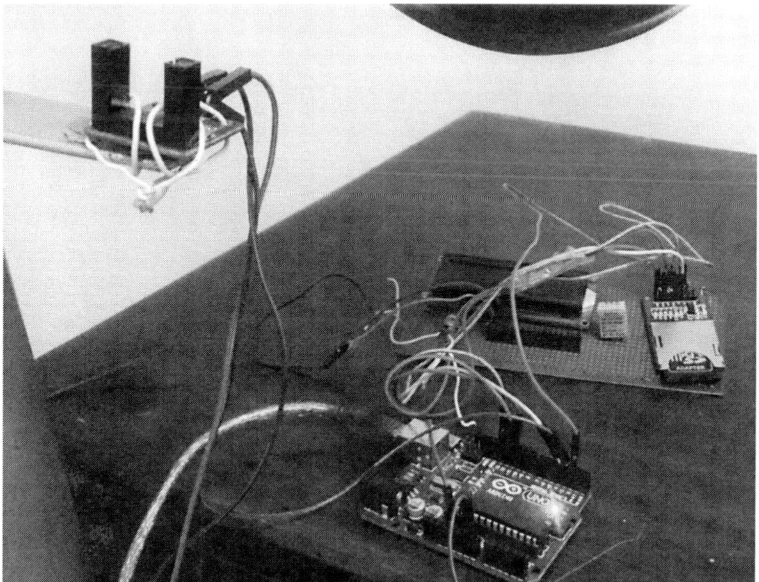

**Figure 5.** Experimental setup.

**Figure 6.** Experimental setup.

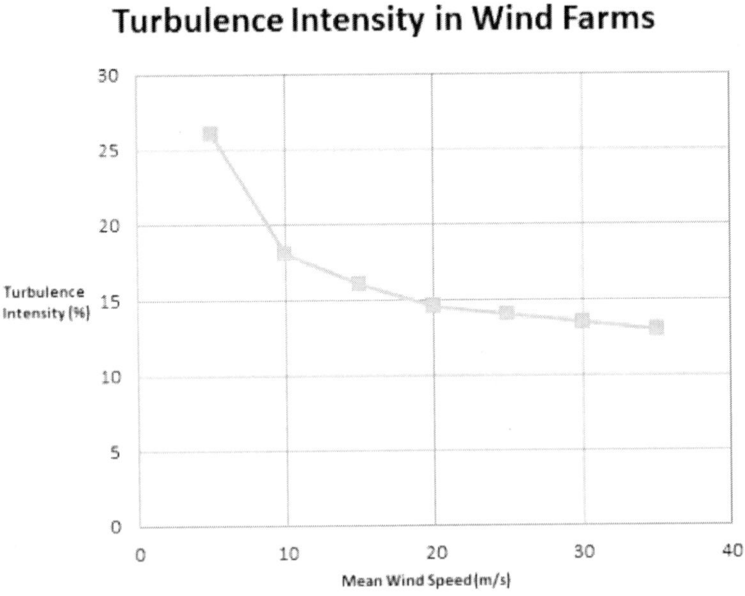


**Figure 7.** Mean wind speed *vs.* turbulence intensity.

## 5.3. Comparison of Simulation and Experimental Results

Figures 8 and 9 show the results of both simulation and experimental model. In Figure 8, overspeeding is evaluated with changing turbulence intensity while in Figure 9 torque is analyzed with changing wind speed. Both figures clearly show that actual model under performs as compared to the simulated model. Here performance refers to the level of accuracy of readings taken by the nodes. The readings refer to measurements of the parameters (discussed in Section 3.1) by the nodes. Figures 8 and 9 also show that simulation results lead to an over prediction of over speeding and torque in anemometers. Figures 8 and 9 also show that the simulation and experimental results show a maximum difference of 10%. The difference in results is basically due to the fact that in actual implementation different other factors (like radio range issues and environmental effects) are also involved which affects the performance of the model *i.e.*, the conditions are not ideal while in simulation, such issues are ignored and the conditions are ideal. The best way to deal with this issue is to consider improving the actual implemented model so that the results can be more accurate instead of working on improving the simulator. By implementing a WSN, the faults in the implemented model are detected and corrected which increases the accuracy of the implemented model. As the number of nodes in our implementation was three so the model performed properly and no faulty node was detected.

As mentioned earlier in the paper, the performance level of implemented projects is 60% of the simulated ones. We compared our simulation and experimental results and came to the conclusion that the accuracy level of implemented projects can be increased upto 86% by accurate assessment and fault detection through a network of sensor nodes. We prove the 86% accuracy level by comparing the simulation results with the experimental results.

# 6. CONCLUSIONS AND FUTURE WORK

In this paper, we have proposed a technique aimed at accurate monitoring of anemometers and detection of their faults by the use of a

WSN. Our simulation and experimental results show that the difference in accuracy of the simulated and actual model is around 10%–15%, which is low as compared to other implementations nowadays that show as low as 60% of the predicted performance *i.e.*, the difference is 40% between them. To the best of our knowledge this is the first paper towards accurate monitoring and fault detection in anemometers for site assessment for improved performance in implemented models.

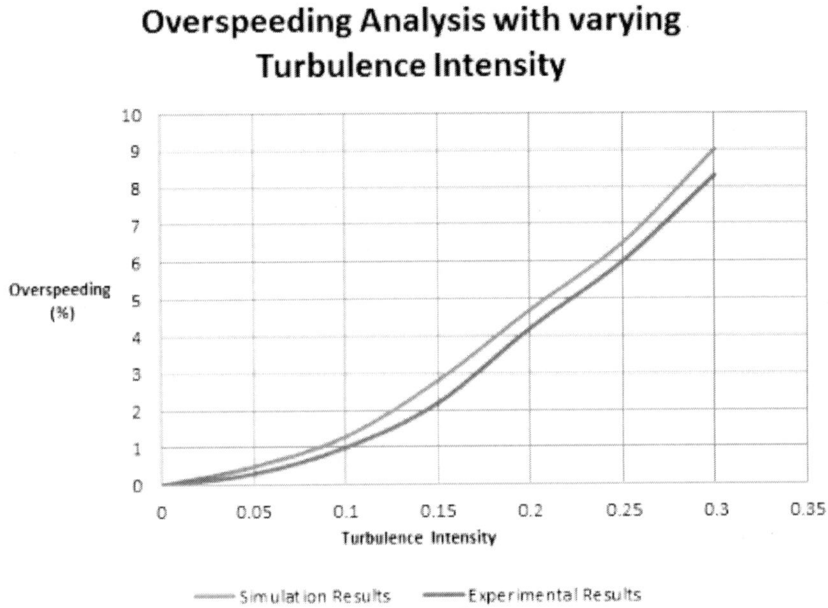


**Figure 8.** Turbulence intensity *vs.* overspeeding.

The possible future work is to implement the proposed technique using more nodes in real hardware for the analysis of a complex network. If the experimental setup is done for a large number of nodes then work can be done on finding the probability of faulty nodes in the network. As this application is less explored, further research in this area can help the wind energy community achieve improved power performance and lower cost of wind technology and allow more widespread applications.

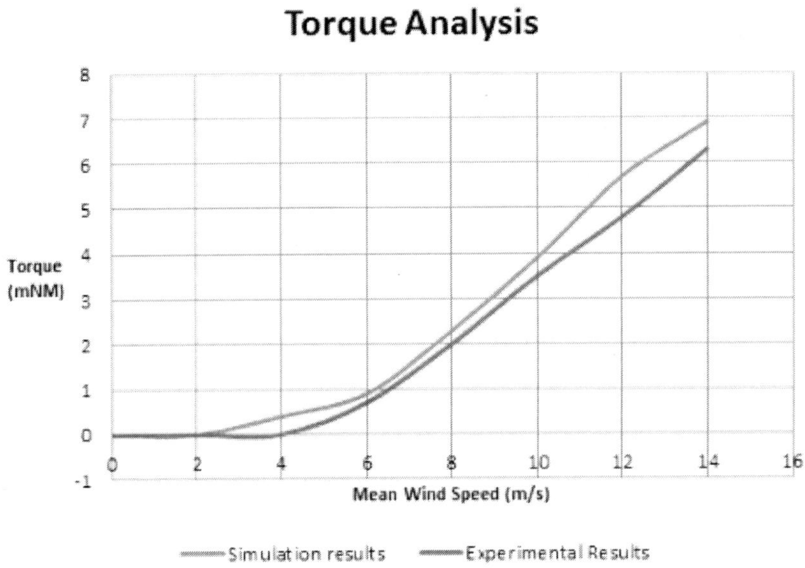


**Figure 9.** Mean wind speed *vs.* torque.

## ACKNOWLEDGMENTS

We are thankful to the anonymous reviewers for giving us their precious time and useful suggestions to improve the overall quality of this paper.

## AUTHOR CONTRIBUTIONS

Komal Saifullah Khan and Muhammad Tariq conceived and designed the experiments; Komal Saifullah Khan performed the experiments; Komal Saifullah Khan and Muhammad Tariq analyzed the data; Muhammad Tariq contributed reagents/materials/analysis tools; Komal Saifullah Khan wrote the paper.

## REFERENCES


1. Gould, I.B. *Wind Energy Deployment in Isolated Islanded Power Systems: Challenges and Realities*; National Renewable Energy Laboratory (NREL): Golden, CO, USA, 2014.

2. Dahlberg, J.A.; Gustavsson, J.; Ronsten, G.; Pedersen, T.F.; Paulsen, U.S.; Westermann, D. Development of a Standardised Cup Anemometer Suited to Wind Energy Applications, Classcup. In *Wind Energy for the New Millennium*; Helm, P., Zervos, A., Eds.; WIP Renewable Energies: München, Germany, 2001; pp. 250–254.

3. Kristensen, L.; Jensen, G.; Hansen, A.; Kirkegaard, P. *Field Calibration of Cup Anemometers*; Risø National Laboratory: Roskilde, Denmark, 2001.

4. World Wind Energy Association. World Wind Energy Report 2008. Proceedings of the 8th World Wind Conference and Exhibition Wind Power for Islands—Offshore and Onshore, Jeju, Korea, 23–25 June 2009.

5. Staddon, J.; Balfanz, D.; Durfee, G. Efficient Tracing of Failed Nodes in Sensor Networks. Proceedings of the 1st ACM International Workshop on Wireless Sensor Networks and Applications 2002, Atlanta, GA USA, 28 September 2002.

6. Chintalapudi, K.K.; Govindan, R. Localized Edge Detection in Sensor Fields. *Ad Hoc Netw.* **2003**, *1*, 273–291.

7. Clouqueur, T.; Saluja, K.K.; Ramanathan, P. Fault Tolerance in Collaborative Sensor Networks for Target Detection.*IEEE Trans. Comput.* **2004**, *53*, 320–333.

8. Krishnamachari, B.; Iyengar, S. Distributed Bayesian Algorithms for Fault-Tolerant Event Region Detection in Wireless Sensor Networks. *IEEE Trans. Comput.* **2004**, *53*, 241–250.

9. Zaher, A.; McArthur, S.D.J.; Infield, D.G.; Patel, Y. Online wind turbine fault detection through automated SCADA data analysis. *Wind Energy* **2009**, *12*, 574–593.

10. Hameed, Z.; Hong, Y.S.; Cho, Y.M.; Ahn, S.H.; Song, C.K. Condition Monitoring and Fault Detection of Wind Turbines and Related Algorithms: A Review. *Renew. Sustain. Energy Rev.* **2009**, *13*, 1–39.

11. Ingelrest, F.; Barrenetxea, G.; Schaefer, G.; Vetterli, M.; Couach, O.; Parlange, M. SensorScope: Application-specific sensor network for environmental monitoring. *ACM Trans. Sens. Netw.* **2010**.

12. Weimer, M.A.; Paing, T.S.; Zane, R.A. Remote Area Wind Energy Harvesting for Low-Power Autonomous Sensors. Proceedings of the 37th IEEE Power Electronics Specialists Conference, 2006 (PESC '06), Jeju, Korea, 18–22 June 2006.

13. Melcher, J.R. *Continuum Electromechanics*; MIT Press: Cambridge, MA, USA, 1981.

14. Teager, H. Some Observations on Oral Air Flow during Phonation. *IEEE Trans. Acoust. Speech Signal Process* **1980**, *28*, 599–601.

15. Andy, S. A Measurement Breeze. *Quality Manufacturing Today* **2009**, 18–20.

16. Haaren, R.V.; Fthenakis, V. GIS-based wind farm site selection using spatial multi-criteria analysis (SMCA): Evaluating the case for New York State. *Renew. Sustain. Energy Rev.* **2011**, *15*, 3332–3340.

17. Longatt, F.G.; Gonzalez, J.S.; Payan, M.B.; Santos, J.M.R. Wind resource atlas of Venezuela based on on-site anemometry observation. *Renew. Sustain. Energy Rev.* **2014**, *39*, 898–911.

18. Wharton, S.; Qualley, G.; Newman, J.; Miller, W. *LIDAR Campaign at Buena Vista Wind Farm: An Examination of Hill Speedup Flows*; Technical Report LLNL-TR-644242; Lawrence Livermore National Laboratory (LLNL): Livermore, CA, USA, 2013.

19. Steven, L.; Eamon, M. LIDAR and SODAR Measurements of Wind Speed and Direction in Upland Terrain for Wind Energy Purposes. *Remote Sens.* **2011**, *3*, 1871–1901.

20. Pedersen, T.F. *Development of a Classification System for Cup Anemometers—CLASSCUP*; Risø National Laboratory: Roskilde, Denmark, 2003.

21. International Electrotechnical Commission (IEC), *Wind Turbines—Part 12-1: Power Performance Measurements of Electricity Producing Wind Turbines*, 1st ed.; IEC: Geneva, Switzerland, 2005.

22. Kristensen, L. *The Cup Anemometer and Other Exciting Instruments*; Risø National Laboratory: Roskilde, Denmark, 1993.

23. Arduino Uno. Available online: http://arduino.cc/en/ Main/arduino BoardUno (accessed on 27 December 2013).

24. Aeolus SimWindFarm. Available online: http://www.ict-aeolus.eu/Sim WindFarm/ (accessed on 1 December 2013).

25. Tariq, M.; Kim, Y.P.; Kim, J.H.; Park, Y.J. Energy Efficient and Reliable Routing Scheme for Wireless Sensor Networks. Proceedings of the International Conference Communication Software and Networks (ICCSN '09), Macau, China, 27–28 February 2009.

26. Cheng, X.; Thaeler, A.; Xue, G.; Chen, D. TPS: A Time-Based Positioning Scheme for Outdoor Sensor Networks. Proceedings of the Twenty-third AnnualJoint Conference of the IEEE Computer and Communications Societies, INFOCOM, Hong Kong, China, 7–11 March 2004.

27. Ruml, W.; Shang, Y.; Zhang, Y. Location from Mere Connectivity. Proceedings of the 4th ACM International Symposium on Mobile Ad Hoc Networking and Computing (MobiHOC '03), Annapolis, MD, USA, 1–3 June 2003; pp. 201–212.

28. Wu, H.; Wang, C.; Tzeng, N.F. Novel Self-Configurable Positioning Technique for Multi-Hop Wireless Networks. *IEEE ACM Trans. Netw.* **2005**, *13*, 609–621.

29. *Wind Turbine Generator Systems-Part 1: Safety Requirements*, 2nd ed.; IEC Technical Committee: Geneva, Switzerland, 1998

# INDEX